Springer Tracts on Transportation and Traffic

Volume 9

Series editor

Roger P. Roess, New York University Polytechnic School of Engineering,
New York, USA
e-mail: rpr246@nyu.edu

About this Series

The book series "Springer Tracts on Transportation and Traffic" (STTT) publishes current and historical insights and new developments in the fields of Transportation and Traffic research. The intent is to cover all the technical contents, applications, and multidisciplinary aspects of Transportation and Traffic, as well as the methodologies behind them. The objective of the book series is to publish monographs, handbooks, selected contributions from specialized conferences and workshops, and textbooks, rapidly and informally but with a high quality. The STTT book series is intended to cover both the state-of-the-art and recent developments, hence leading to deeper insight and understanding in Transportation and Traffic Engineering. The series provides valuable references for researchers, engineering practitioners, graduate students and communicates new findings to a large interdisciplinary audience.

More information about this series at http://www.springer.com/series/11059

Alp Caner · Polat Gülkan · Khaled Mahmoud
Editors

Developments in International Bridge Engineering

Selected Papers from Istanbul Bridge Conference 2014

Springer

Editors
Alp Caner
Civil Engineering Department ODTU
Middle East Technical University
Ankara
Turkey

Khaled Mahmoud
Bridge Technology Consulting
New York, NY
USA

Polat Gülkan
Department of Civil Engineering
Cankaya University
Ankara
Turkey

ISSN 2194-8119 ISSN 2194-8127 (electronic)
Springer Tracts on Transportation and Traffic
ISBN 978-3-319-19784-5 ISBN 978-3-319-19785-2 (eBook)
DOI 10.1007/978-3-319-19785-2

Library of Congress Control Number: 2015943363

Springer Cham Heidelberg New York Dordrecht London

Printed on acid-free paper

Springer International Publishing AG Switzerland is part of Springer Science+Business Media (www.springer.com)

Preface

The movement of people, goods, and services between urban settlements and production facilities is the essential circulation that keeps regional and national economies running. Bridges are the vital connectors that ensure that transportation systems function safely and economically as intended by those who have planned and conceptualized them. Bridge engineering is a special discipline within structural engineering that focuses on the design, construction, erection, assessment, and maintenance of these connectors for the benefit of society. Bridges are often durable symbols of technical prowess, economic power, and esthetic refinement. Just as many ancient bridges that have become landmarks evoke in us a sense of amazement at the creative skills of the old masters who built them, so do their younger counterparts that display the flair of more recent builders who have combined advanced analysis techniques with modern, high-performance materials to craft structures that defy the heretofore unchallenged limits in terms of span, height, or frugal use of material resources. Its many counterexamples notwithstanding, bridge engineering is indeed a uniquely singular parade arena to display technical experience and artistic maturity. It is sometimes a shared responsibility among engineers and the special breed of architects who have the craftsmanship to improve the technical attributes of the finished product. We hope that this book will serve to give its readers an overview of the many problems that must be resolved to produce bridge structures that will meet the challenges they will face during their service lifetime.

The Turkish Group of the International Association for Bridge and Structural Engineering (IABSE), in collaboration with Middle East Technical University (METU), the Turkish General Directorate for State Highways (KGM), the General Directorate for State Railways (TCDD), the Turkish Contractors Association, the Turkish Association for Structural Steel (TUCSA), the Turkish National Committee for Roads (YTMK), the Turkish Association for Seismic Isolation, the International Association for Earthquake Engineering (IAEE), the Transportation Research Board (TRB), American Concrete Institute (ACI), the Bridge Engineering Association, and SIBERC served as host for the Istanbul Bridge Conference during August 11–13, 2014. The conference was generously supported by many sponsors, AGM, Arsan

Kaucuk, Aydiner Inc. Company (Aybet), BERD, Besmak, Cengiz Nurol, Chodai Co., Computers and Structures, Inc. (CSI), DOKA, DUYGU Engineering, EMAY International Engineering and Consultancy Inc, EM-KE, Endem Construction, EPO Construction Chemicals, FIP Industriale, FREYSAŞ, IC İçtaş Construction Co, Ilgaz İnşaat, INPRO, JOTUN, KMG Project, KOBA Engineering and Consulting Co. Ltd, LARSA, MAGEBA, Mapa İnşaat ve Ticaret A.Ş., Maurer Söhne GmbH & Co., MEGA Engineering Consulting Co., Mistras Group, MOOG Gmbh, Otoyol Yatırım ve İşletme A.Ş., Ozdekan Rubber Company, PERİ, Pitchmastic Pmb Ltd, Sismolab, Strainstall, TEMELSU International Engineering Services, TESTART, TNO Diana, TTS International Engineering and Architecture, WireCo Structures, Wowjoint, Wuhan HIRUN Engineering Equipment Co., YAPIFEN Engineering, and Yüksel Proje.

iBridge was attended by **250** participants representing **30** different countries. The conference program included **96** papers. It was complemented by a visit to the construction site for the third Istanbul Strait crossing, a major project that will place the bridge among the top five long-span bridges in the world in several categories when it has been completed.

The 19 papers selected for this book represent a cross section of the breadth of the topics that were taken up in the conference. The selection has been made on the basis of scientific and technical relevance, timeliness, and expected added value for practitioners. A few authors who were invited to finalize their papers for inclusion in the book failed to do so on time, and a few others were eliminated because they had been submitted elsewhere for publication. The papers represent current concerns in bridge engineering as they exist in many countries and the solutions for those issues that have been arrived at under codified constraints that have echoes in other locations. Theory, experiment, and practice are presented in balanced proportions in the texts that follow.

We have grouped them under four broadly defined thematic areas as follows:

- "Modeling and Analysis" (7 papers)
- "Construction and Erection Techniques" (3 papers)
- "Design for Extreme Events" (4 papers)
- "Condition Assessment and Structural Health Monitoring" (5 papers)

The book has been divided into four principal sections along these themes.

The editors would like to acknowledge the excellent cooperation that they have received from the authors in revising their papers according to reviewer comments and the tens of reviewers, from the Scientific Committee as well as externally named experts (many in multiple assignments), who willingly gave much of their time in examining the papers for their content to ensure that this would be a printed source with a positive impact, and provide guidance for their colleagues in the broader engineering community. The editorial staff at Springer Publishers has been unfailingly helpful in guiding us toward producing a book in their style.

We have received invaluable support from the staff at INTERCON that served as the Conference Secretariat and doubled as providers for the initial screening process of the written contributions.

Alp Caner
Polat Gülkan
Khaled Mahmoud

Contents

Part II Construction and Erection Techniques

Part III Design for Extreme Events

Part IV Condition Assessment and Structural Health Monitoring

Part I
Modeling and Analysis

Optimal Design of Pile Foundation in Fully Integral Abutment Bridge

Jun Qing Xue, Bruno Briseghella, Bao Chun Chen, Pei Quan Zhang and Tobia Zordan

Abstract In order to resolve the durability problem of expansion joints and bearings, the integral abutment bridge (IAB) has become more and more popular. For integral abutment bridge, choosing a suitable pile foundation type is a challenging problem, because the substructure is fixed with superstructure to bear the load together. In this paper, the design of the pile foundation in a fully integral abutment bridge (FIAB) in China was analyzed. A finite element model was built by the commercial software MIDAS considering soil-structure interaction and construction stage simulation. A sensitive analysis was carried out to investigate the influence of different pile foundation types on the mechanic performance of the IAB. The results show that when the circular pile is used, the stress of pile, negative moment and tensile stress of girder are smaller than those when the rectangular pile is used. With the increase of pile diameter, the stress and displacement of pile decrease, while the bending moment of pile and the negative moment and tensile stress of girder increase. For rectangular piles, with the increase of cross-sectional length-width ratio, the bending moment of pile, negative moment and tensile stress of girder decrease; while the stress and displacement of pile increase.

1 Introduction

At present, many jointed bridges have been constructed all over the world, in which expansion joints and bearings are installed in order to absorb cyclic thermal expansion and contraction, creep and shrinkage, and differential settlement. The expansion joints and bearing of bridges are easily damaged during service life of

J.Q. Xue (✉) · B. Briseghella · B.C. Chen · P.Q. Zhang · T. Zordan
College of Civil Engineering, Fuzhou University, 2 Xue Yuan Road, University Town, Fuzhou 350108, Fujian, People's Republic of China
e-mail: junqing.xue@fzu.edu.cn

A. Caner et al. (eds.), *Developments in International Bridge Engineering*,
Springer Tracts on Transportation and Traffic 9,
DOI 10.1007/978-3-319-19785-2_1

many jointed bridges due to the influence of increasing age, climate, environment and some unpredictable destructive effects of nature [1]. Furthermore, the damaged expansion joints and bearings will produce a series of secondary diseases. Moreover, the maintenance cost and vehicle restrictions caused by frequent repair or the serious social impact are remarkable [2, 3]. In order to resolve the durability problem of expansion joints and bearings, many engineers proposed the concept of “No expansion joint is the best joint”. In this case, the integral abutment bridge (IAB), which can eliminate expansion joints and bearings, becomes a challenging solution [4, 5]. Different with jointed bridges, the abutments of integral abutment bridges are fixed to the girders. The superstructure, abutments and pile foundations, backfill and soil around pile foundations work as a frame. Therefore, the type of pile foundation has a great impact on the mechanical performance of integral abutment bridges, which can be considered as one of the key problems in the design of IAB.

2 Case Study

In this paper, a fully integral abutment bridge (FIAB) in China was chosen as case study. The bridge is located in Shi-An expressway. The bridge is a two-span prestressed concrete fully integral abutment bridge. The main features are listed in Table 1. The elevation layout and typical cross section are shown in Figs. 1 and 2, respectively. The details of fully integral abutment and approach slab are illustrated in Fig. 3.

Table 1 Main features of bridge

Total length	41.8 m	
Number of spans	2	
Girder information	Number per span	4
	Height	1.2 m
	Type	Box-girder
Deck information	One carriageway	12.5 m × 0.22 m
Pier height (cap + column)	1.4 + 5.0 m	
Abutment height (backwall + stem)	1.3 + 1.4 m	
Pile	Number per each abutment	7
	Cross-sectional shape	Circular
	Diameter	0.7 m
	Length	34 m
	Number per each pier	1
	Cross-sectional shape	Circular
	Diameter	1.5 m
	Length	45 m

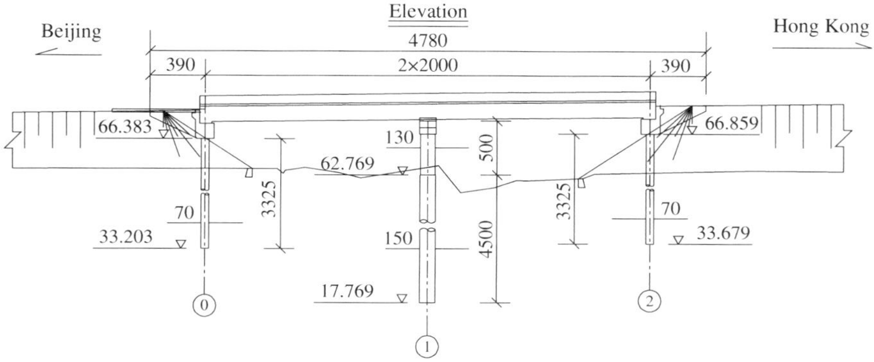

Fig. 1 Elevation layout (cm)

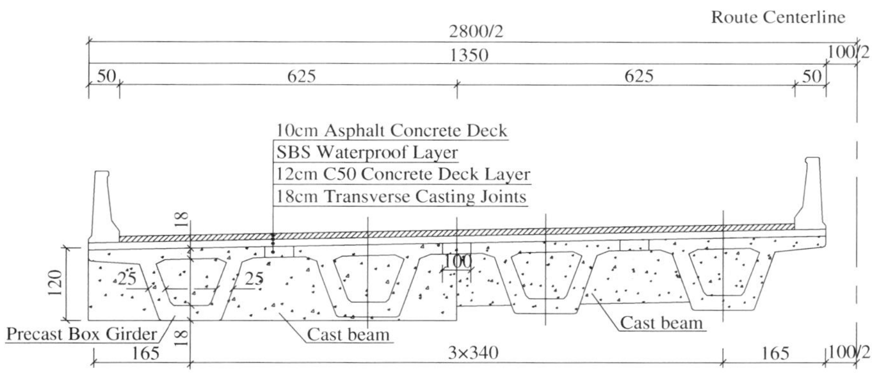

Fig. 2 Typical cross section of precast box girder (cm)

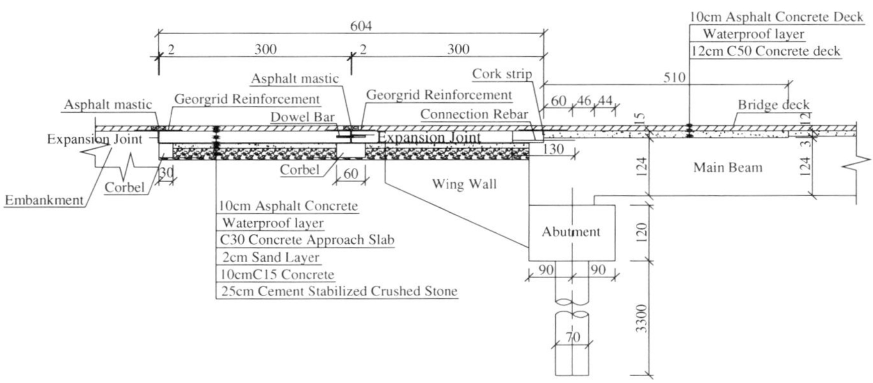

Fig. 3 Details of fully integral abutment and approach slab (cm)

3 Finite Element Model

In this paper, a 3D finite element model was built by commercial finite element software MIDAS/Civil2012. The superstructure was simulated by grillage method, in which the longitudinal beam elements and virtual transverse beam elements were used to simulate the longitudinal girders and wet joint connections between longitudinal girders. The height of virtual transverse beam elements is set as the same as the upper plate height of longitudinal girders. Different with the jointed bridge, of which the substructure modeling can be neglected and simplified as the simply supported boundary condition, the substructure of fully integral abutment bridge, including piers, abutments and piles, should be considered in the model, because the substructure-soil interaction can not be ignored. In this case, the beam element was used to simulate the substructure. The bearings installed on the top of pier cap were simulated by elastic connections. For the fully integral abutment bridge, the superstructure-abutment connections were modeled by rigid connections. The finite element model is shown in Fig. 4.

In order to accurately simulate the performance of bridge, four construction stages were considered in the model, because the construction technology of simply supported-continuous was used in the real case.

1. Simply Supported Stage: The substructure and girders are activated. The temporary supports were used as the connections between girders and abutments or girders and piers. The dead load and prestressing force which will produce positive bending moment are considered.
2. Continuous Stage: The diaphragms, wet joints and other continuous segments are activated. The rigid connection is applied between girders and abutments.

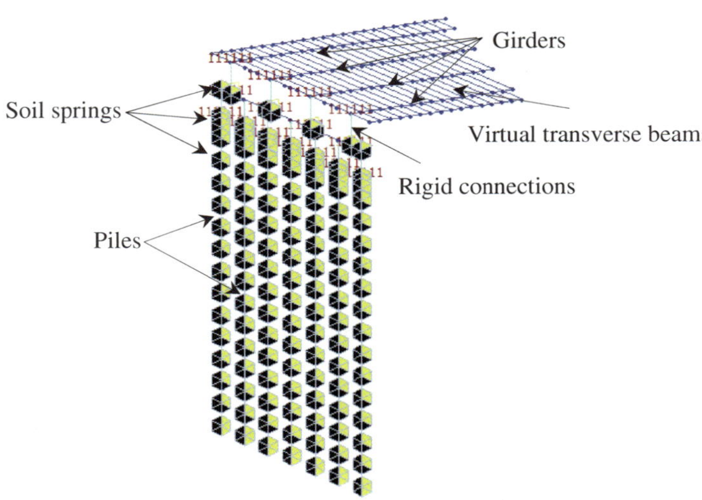

Fig. 4 Finite element model of bridge

The super dead load and prestressing force which will produce negative bending moment are considered.
3. Shrinkage and Creep Stage: The shrinkage and creep for 10 years (3650 days) are considered.
4. Completion Stage: The uniform temperature raising 34 °C or falling 10 °C, the positive and negative temperature gradient, vehicle load, soil pressure on abutments and pile-soil interaction are considered.

According to "General Code for Design of Highway Bridges and Culverts" (JTG D60-2004) [6], some load combinations should be considered, including ultimate limit state combinations I, II, III and serviceability limit state combinations for short-term action effects (combination IV), as listed in following.

1. Combination I: 1.2 dead load + 1.0 prestress + 1.0 shrinkage and creep + 1.4 vehicle load (including impact);
2. Combination II: 1.2 dead load + 1.0 prestress + 1.0 shrinkage and creep + 1.4 vehicle load (including impact) + 0.8 × 1.4 × uniform temperature;
3. Combination III: 1.2 dead load + 1.0 prestress + 1.0 shrinkage and creep + 1.4 vehicle load (including impact) + 0.7 × 1.4 × (uniform temperature + temperature gradient).
4. Combination IV: 1.0 dead load + 1.0 prestress + 1.0 shrinkage and creep + 0.7 vehicle load (excluding impact) + 1.0 uniform temperature + 0.8 temperature gradient.

4 Substructure-Soil Interaction

Different with jointed bridges, the interaction between soil and substructure should be considered for fully integral abutment bridges, including the pile-soil interaction and abutment-backfill interaction. In this paper, the substructure-soil interactions were simulated by spring elements, of which the stiffness was calculated by "m" method proposed by Chinese code "Code for Design of Ground Base and Foundation of Highway Bridge and Culverts" (JTG D63-2007) [7]. According to geologic survey report, two soil layers, including silt and silty clay, could be found at bridge site. The piles should be designed as the friction pile, because the bearing capacity of soil is low. For the backfill, the artificial fill was used. In this paper, the substructure-soil interaction was simulated based on Appendix P of Chinese code "Code for Design of Ground Base and Foundation of Highway Bridge and Culverts" (JTG D63-2007) [7]. Based on Winkler's Hypothesis, the lateral soil resistance σ_{zx} at a certain depth z is proportional to the horizontal deflection X_z, as expressed by Eq. 1. According to the "m" method proposed in Chinese code [7], the foundation coefficient can be calculated by using Eq. 2. Based on the theory of elastic foundation beam, the piles can be considered as the girders supported by a series of soil springs. The finite element model of pile-soil interaction below ground line is shown in Fig. 5. In this paper, the the value of "m" of soil layers and backfill are all 5000. The calculation process of

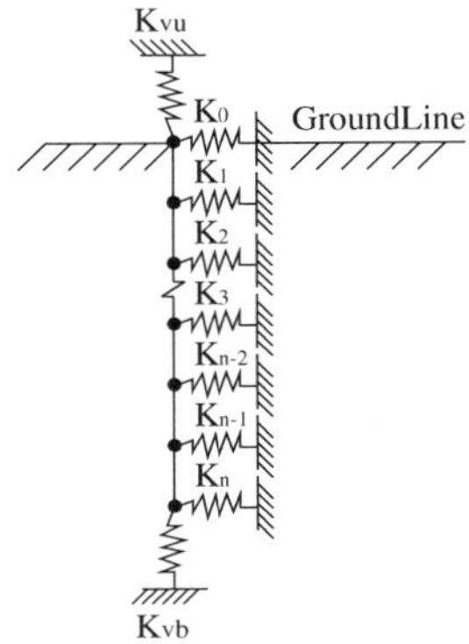

Fig. 5 Finite element model of pile-soil interaction

the horizontal soil spring stiffness K_Z is introduced in following. For each horizontal soil spring, the stiffness can be calculated by Eq. 3, which expresses the relationship between the restoring force F_z and lateral deflection of pile at z depth X_z. Equation 2 can be substituted into Eq. 1 to obtain Eq. 4. According to the relationship between force and reacting force, the restoring force F_z can be calculated by Eq. 5. Comparing Eq. 3 and Eq.5, Eq. 6 can be obtained.

$$\sigma_{ZX} = X_Z C_Z \tag{1}$$

$$C_Z = mz \tag{2}$$

$$F_Z = K_Z X_Z \tag{3}$$

$$\sigma_{ZX} = mzX_Z \tag{4}$$

$$F_Z = R_Z A_Z = mzX_Z b_1 h_Z \tag{5}$$

$$K_Z = mzb_1 h_Z \tag{6}$$

where, C_Z is the foundation coefficient; m is foundation coefficient of soil varied with depth (kN/m^4); K_z is the stiffness of the soil spring at z depth; A_z is the action area of soil pressure at z depth; h_z is the thickness of soil at z depth; b_1 is the calculating width of pile, which is 0.95 m in this case, according to the Chinese code [6]. Due to the space limitation, the calculation process of the calculating width is neglected.

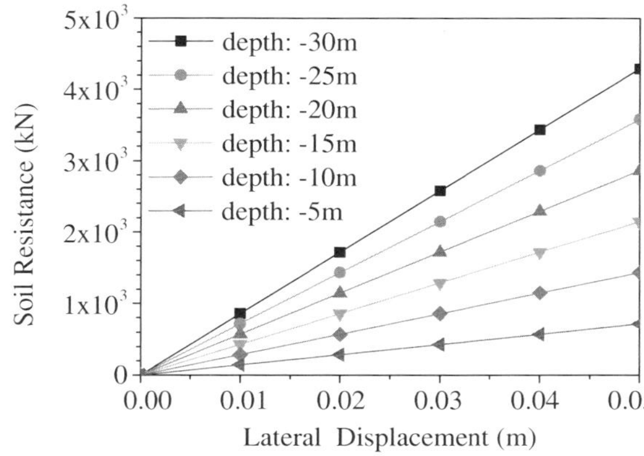

Fig. 6 Soil resistance-pile lateral displacement relationship

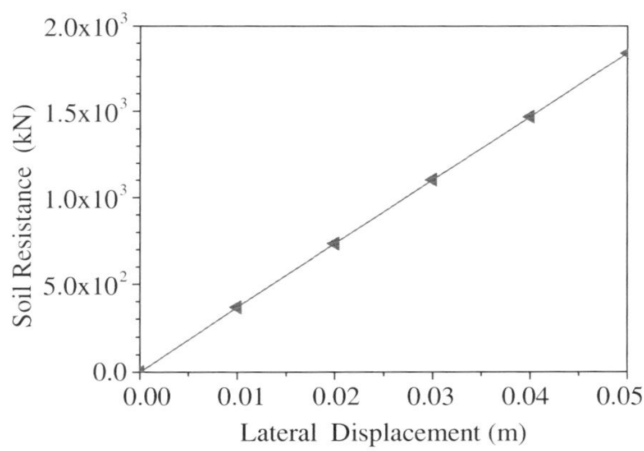

Fig. 7 Passive soil resistance-abutment lateral displacement relationship

The stiffness curves of horizontal springs at several depths of piles and abutments are shown respectively in Figs. 6 and 7.

5 Optimal Analysis of Pile Foundation Type

As introduced above, the piles should be designed as friction piles according to the geological data. Therefore, the vertical friction of all piles should remain constant during analysis, which means that the vertical area of all piles can not be changed.

The checking calculation of piles and girders according to "Code for Design of Highway Reinforced Concrete and Prestressed Concrete Bridges and Culverts" (JTGD 62-2004) [8] are introduced in the following.

Considering load combinations I, II and III, the maximum horizontal displacements and internal forces of piles can be considered as the most unfavorable working conditions in the ultimate limit state. For the strength checking of pile, the pile strength in the serviceability limit state should be considered. The eccentric compression/tension, when the primary moment is the maximum (Eccentric-Mymax) and when the primacy moment is the minimum (Eccentric-Mymin) should be checked. For each section of pile, the most unfavorable load combination should be used respectively. The compressive and tensile capacity of piles can be calculated according to Chinese code (JTG D62-2004) [8].

According to Chinese code (JTG D62-2004) [8], the ultimate stress of the precast member (A-class prestressed concrete) in the serviceability limit state combination for short-term action effects (combination IV) should be checked by Eq. 7.

$$\sigma_{tp} \leq 0.7 f_{tk} \tag{7}$$

where, σ_{tp} is the main concrete tensile stress; f_{tk} is the standard value of concrete tensile strength. In this case, the concrete grade of girder is C50 ($0.7f_{tk} = 1.855$ MPa).

5.1 *Influence of Pile Cross-sectional Shape*

When the number and vertical area of piles are fixed, four types of the pile cross-sectional shape could be considered, including (1) Circular cross section (Φ70 cm); (2) Rectangular cross section (55 cm × 55 cm); (3) Rectangular cross section (65 cm × 45 cm); (4) Rectangular cross section (70 cm × 40 cm). The layouts of pile foundations with different cross-sectional shapes are shown in Fig. 8. The influence of different cross-sectional shapes of piles on the internal forces and deformations of piles and girders are compared in Tables 2 and 3. It could be found that when the circular pile is used, the stress of pile, negative moment and tensile stress of girder are smaller than those when the rectangular pile is used. For rectangular piles, with the increase of cross-sectional length-width ratio, the bending moment of pile, negative moment and tensile stress of girder decrease; while the stress and displacement of pile increase.

The strength checking of piles for different cross-sectional shapes are illustrated in Fig. 9. Due to the space limitation, only the checking of eccentric tension, when the primary moment is the maximum, is shown in this paper. The circles in the figures represent the failure regions, where the pressure design value is larger than the comprehensive bearing capacity. The strength checking results indicate that only the circular pile with the diameter of 70 cm can meet the requirements.

5.2 *Influence of Pile Diameter and Number of Piles*

When the pile cross-sectional shape is fixed as circular and the cross-sectional reinforcement ratio keeps invariant, four different types of pile diameters were chosen as parameters, including (1) 70 cm; (2) 80 cm; (3) 100 cm; (4) 120 cm. As mentioned above, the vertical area of pile foundations is fixed. Therefore, the number of piles should vary corresponding to the variation of the pile diameters, which are (1) 7; (2) 6; (3) 5; (4) 4. The layouts of pile foundations with different pile diameters and

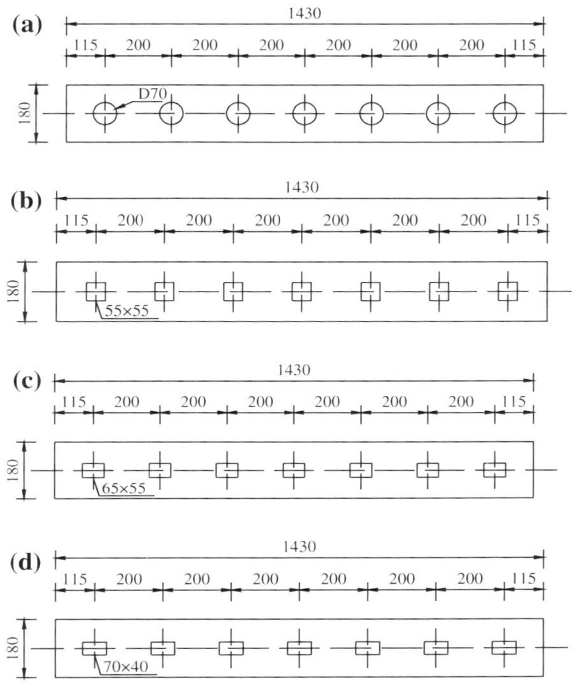

Fig. 8 Layouts of pile foundations with different cross-sectional shapes (cm). **a** Circular cross section (Φ70 cm). **b** Rectangular cross section (55 cm × 55 cm). **c** Rectangular cross section (65 cm × 45 cm). **d** Rectangular cross section (70 cm × 40 cm)

Table 2 Internal forces and deformations of piles for different pile cross-sectional shapes

Pile type	Maximum bending moment (kN m)	Compare to Φ70 (%)	Maxi-stress (MPa)	Compare to Φ70 (%)	Pile dis-placement (m)	Compare to Φ70 (%)
Φ70	388.7	100.0	11.4	100.0	0.0082	100.0
55 × 55	405.4	104.3	16.5	144.7	0.0076	92.7
65 × 45	361.6	93.0	19.5	171.1	0.0077	93.9
70 × 40	332.7	85.6	21	184.2	0.0078	95.1

Table 3 Internal forces of girders adjacent to abutments for different pile cross-sectional shapes

Pile type	Maximum negative moment (kN m)	Compare to Φ70 (%)	Maximum tensile stress (MPa)	Compare to Φ70 (%)
Φ70	1167.3	100.0	1.30	100.00
55 × 55	1612.9	138.2	1.42	109.23
65 × 45	1538.8	131.8	1.40	107.69
70 × 40	1480.2	126.8	1.37	105.38

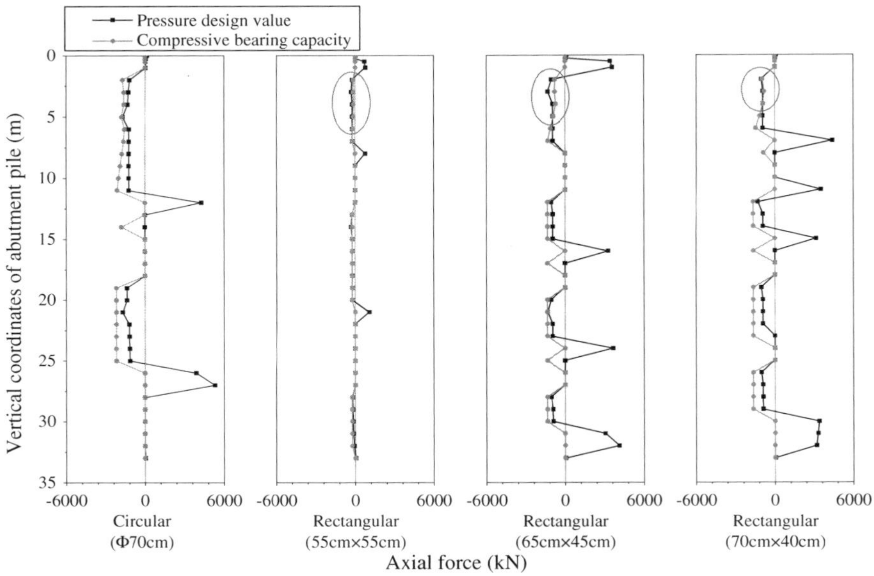

Fig. 9 Checking of eccentric tension-Mymax of piles for different pile cross-sectional shapes

Fig. 10 Layouts of pile foundations with different pile diameter and number of piles (cm). **a** Circular cross section (7Φ70 cm). **b** Circular cross section (6Φ80 cm). **c** Circular cross section (5Φ100 cm). **d** Circular cross section (4Φ120 cm)

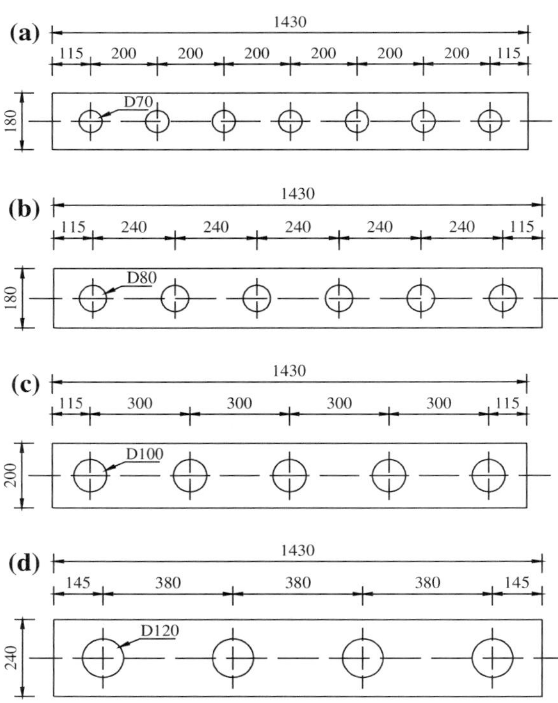

Table 4 Internal forces and deformations of piles for different pile diameters and number of piles

Pile type	Maximum bending moment (kN m)	Compare to Φ70 (%)	Maxi-stress (MPa)	Compare to Φ70 (%)	Pile dis-placement (m)	Compare to Φ70 (%)
7Φ70	388.7	100.0	11.4	100.0	0.0082	100.0
6Φ80	510.4	131.3	9.5	83.3	0.0082	100.0
5Φ100	695.4	178.9	7.5	65.8	0.0081	98.8
4Φ120	973.9	250.6	7.2	63.2	0.0081	98.8

Table 5 Internal forces of girders adjacent to abutments for different pile diameter and number of piles

Pile type	Maximum negative moment (kN m)	Compare to 7Φ70 (%)	Maximum tensile stress (MPa)	Compare to 7Φ70 (%)
7Φ70	1167.3	100.0	1.30	100.00
6Φ80	1226.1	105.0	1.58	121.54
5Φ100	1378.9	118.1	1.69	130.00
4Φ120	1402.6	120.2	1.71	131.54

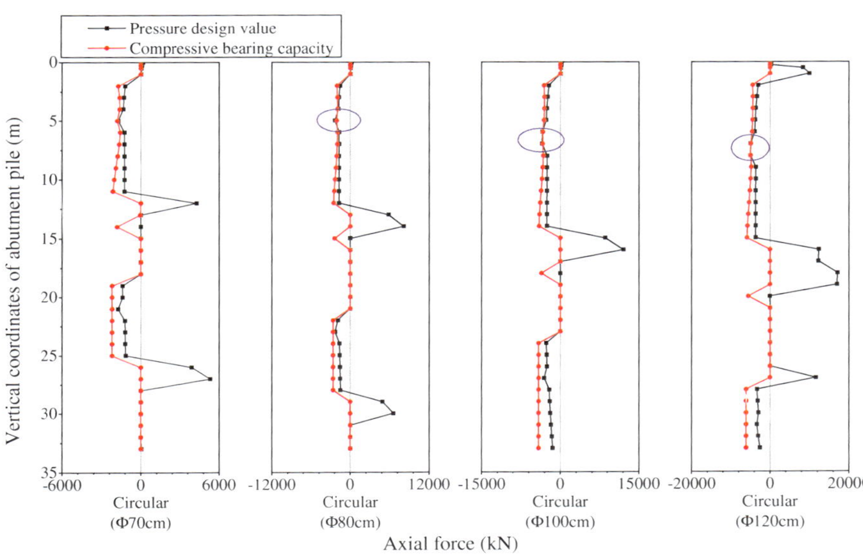

Fig. 11 Checking of eccentric tension-Mymax of piles for different pile diameter and number of piles

number of piles are shown in Fig. 10. The influence of different pile diameters and number of piles on the internal forces and deformations of piles and girders are compared in Tables 4 and 5. It could be found that with the increase of pile diameter, the stress and displacement of pile decrease, while the bending moment of pile and the negative moment and tensile stress of girder increase.

The strength checking of piles for different pile diameter and number of piles are illustrated in Fig. 11. Due to the space limitation, only the checking of eccentric tension, when the primary moment is the maximum, is shown in this paper. The circles in the figures represent the failure regions, where the pressure design value is larger than the comprehensive bearing capacity. It could be found that when the layout of circular piles (7Φ70 cm) is used, the strength checking of piles can meet the requirement. Therefore, the layout of circular piles (7Φ70 cm) is suggested in this case.

6 Conclusions

In this paper, the finite element model of a fully integral abutment bridge (FIAB) in China was built. The influence of different pile foundation types on the mechanic behaviours of bridge were analyzed, including the pile cross-sectional shape, pile diameter and number of piles. Some conclusions obtained for optimal design of piles in fully integral abutment bridge are listed in the following.

1. When the circular pile is used, the stress of pile, negative moment and tensile stress of girder are smaller than those when the rectangular pile is used.
2. For rectangular piles, with the increase of cross-sectional length-width ratio, the bending moment of pile, negative moment and tensile stress of girder decrease; while the stress and displacement of pile increase.
3. With the increase of pile diameter, the stress and displacement of pile decrease, while the bending moment of pile and the negative moment and tensile stress of girder increase.

References

1. Chen BC, Zhuang YZ, Briseghella B (2013) Jointless bridges. In: Book. People's Communications Press, Beijing
2. Zordan T, Briseghella B, Lan C (2011) Analytical equation for limit length of integral abutment bridges. Struct Eng Int 21(3):304–310
3. Zordan T, Briseghella B, Lan C (2011) Parametric and pushover analyses on integral abutment bridge. Eng Struct 33(2):502–515
4. Briseghella B, Xue JQ, Lan C, Zordan T, Chen BC (2014) Maximum length of integral abutment bridges. J Archit Civ Eng 31(1):104–110

5. Xue JQ (2013) Retrofit of existing bridges with concept of integral abutment bridge—static and dynamic parametric analyses. PhD thesis, University of Trento, Italy
6. CCCC Highway Consultants CO., Ltd. (2004) JTG D60-2004 general code for design of highway bridges and culverts. People's Communications Press, Beijing
7. CCCC Highway Consultants CO., Ltd. (2007) JTG D63-2007 code for design of ground base and foundation of highway bridge and culverts. People's Communications Press, Beijing
8. CCCC Highway Consultants CO., Ltd. (2004) JTG D62-2004 code for design of highway reinforced concrete and prestressed concrete bridges and culverts. People's Communications Press, Beijing

Analysis of the Response Under Live Loads of Two New Cable Stayed Bridges Built in Mexico

Roberto Gómez, Raul Sánchez-García, J.A. Escobar and Luis M. Arenas-García

Abstract In this paper we study the response, under live loads, of two new cable stayed bridges (Baluarte and Carrizo), recently open to traffic. Both bridges are located in the northern pacific mountains of Mexico along the Mazatlan-Durango new highway at approximately 50 km from the coast. These bridges happen to be the most important bridge structures of the highway. Both bridges have the same fan cable pattern and similar type of superstructure. However the response of both decks under live loads is quite different as will be shown. This paper describes also the numerical analyses to evaluate the bridges structural response under static and moving loads mainly. From this evaluation and assessment, conclusions and recommendations related to their structural stability are provided. The activities described were performed before opening the bridges and comprised also the instrumentation with fiber optic sensors, live load tests and ambient vibration tests.

Keywords Vehicular loads · Cable-Stayed bridge · Load testing · Analytical response bridge

1 Introduction

Built to cross the Pacific mountains of Mexico (Western Sierra Madre) a new highway was open to traffic last year. The highway has a total length of 230 km and 12 m wide to accommodate 2 lanes. However, from 156 + 300 km to 164 + 000 km the highway has four lanes. Along the highway 63 tunnels and 115 bridges were built. Among the structures of greater importance in this road are, in

R. Gómez (✉) · R. Sánchez-García · J.A. Escobar · L.M. Arenas-García
Institute of Engineering, National Autonomous University of Mexico,
Ciudad Universitaria, Coyoacán, Mexico City, Mexico
e-mail: RgomezM@iingen.unam.mx

A. Caner et al. (eds.), *Developments in International Bridge Engineering*,
Springer Tracts on Transportation and Traffic 9,
DOI 10.1007/978-3-319-19785-2_2

first place, the "Baluarte" cable stayed bridge with a length of 1124 m and a height of 402 m, second, the "Sinaloense" tunnel with 2.78 km length and, third, another cable stayed bridge, the "Carrizo", at 162 + 172 km.

Since the beginning of the construction of the highway, the Ministry of Communications and Transport of Mexico (SCT) commended, among other tasks, to the Institute of Engineering of UNAM (IIUNAM), the instrumentation and monitoring of both bridges in different important aspects and stages of their construction. These activities developed by IIUNAM during the construction of the bridges can be summarized in the following major tasks: monitoring of displacements of rock stratums; tilts and stresses in pylon 5; wind velocities; topography; and instrumentation and load tests of the main span.

In this paper we only describe results of field tests under static loads carried out in the main section of the superstructures. This information was useful to calibrate the mathematical models, as well as to confirm the recommendations of the wind and seismic design studies developed previously to the design stage of the bridge.

In another section of the paper, we present results of the analyses of the bridges under moving loads representing design live loads used in Mexico. Implications of the results in the structural behavior of the bridges are presented in the conclusions.

2 The Bridges

2.1 The "Baluarte" Bridge

This cable-stayed bridge is an icon of Mexican bridge engineering due to its particular features and location. The bridge crosses a ravine of about 400 m depth. It is the longest cable-stayed bridge in North America with a total length of 1124 m: a main composite steel-concrete span of 520 m (the longest ever built in Latin-America) and two post-tensioned concrete lateral spans of 250 and 354 m, respectively. A total 152 stays are arranged in a semi harp configuration in eight vertical planes. A panoramic view of the bridge is shown in Fig. 1.

The two main steel girders of the central span are conformed by welded steel plates. Segments of girders of 12 m length were fabricated in a remote shop and delivered to the site where they were lifted to assemble 12 m segments of superstructure. Once the correspondent two girders and cross frames were assembled, the concrete slab was cast in situ. The central span was erected simultaneously from both pylons. An approximate deck width of 19.76 m is provided to carry four traffic lanes; the lateral spans have two main bodies of post-tensioned concrete box segments with approximately 22 m of deck width; at the vicinity of the pylons on the central span the same concrete box segments were used along 44.0 m; steel cross beams are used in both types of superstructures (Fig. 2).

Pylons have an inverted Y shape and frame type piers are used along the backspans. The substructure is founded on cast in place footings.

Fig. 1 The "Baluarte" bridge, panoramic view

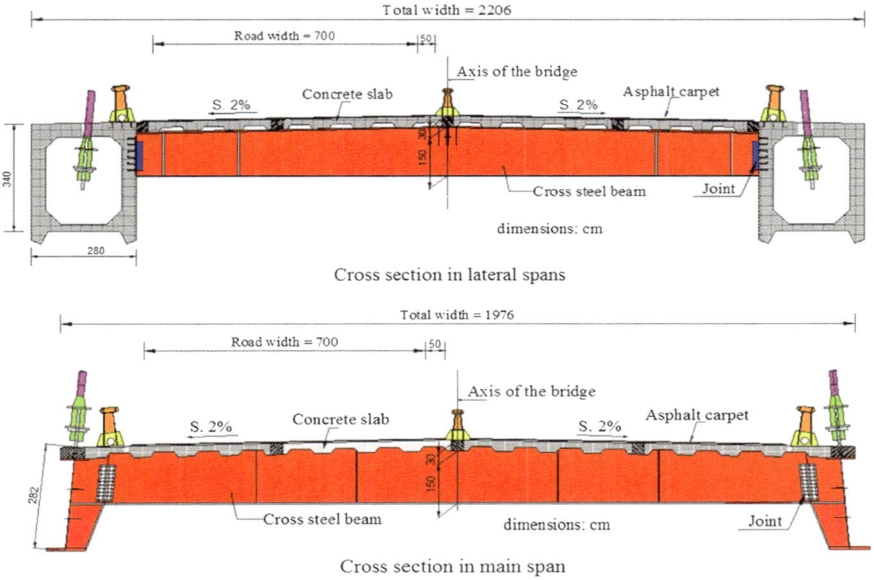

Fig. 2 Cross sections of the superstructure, "Baluarte" bridge

2.2 The "Carrizo" Bridge

This bridge has a total length of 434.6 m with three spans including two cable stayed spans (182 m each) and a double cantilever section (70.6 m). Figure 3 shows a panoramic view from the bridge. Erection of the cable stayed spans was executed by a cantilever method. Stays are installed in two vertical planes in a semi fan layout. The total number of stays is 56.

The cross section of the superstructure is very similar to the one used in the "Baluarte" bridge (Fig. 4); post-tensioned box segments were used for the construction of the double cantilever of 70.6 m, this section of the bridge is joined to the cable stayed section by means of a special joint. The total width of the deck is 18.40 m to accommodate 4 lanes (Fig. 4). The main member of the substructure is pylon 2, it has an H shape and a height of 226 m can be measured from the level of the foundation to its top level; pier number 3 is of the frame type and is 70 m height. Two massive concrete abutments were constructed at the ends of the bridge. A simple supported single span (38 m) connects the "Carrizo" bridge to the portal of an adjacent tunnel.

After completion of the design process [1, 2], wind tunnel specific studies were carried out for each bridge [3, 4]. Performance of the superstructure under different wind loading conditions was evaluated. After these studies, wind fairings were proposed and installed along the main span of the "Baluarte" bridge and along the total length of the metallic section of the "Carrizo" bridge.

The design and construction projects of both bridges were developed by the Ministry of Transportation and Communication of Mexico.

Fig. 3 The "Carrizo" bridge, panoramic view

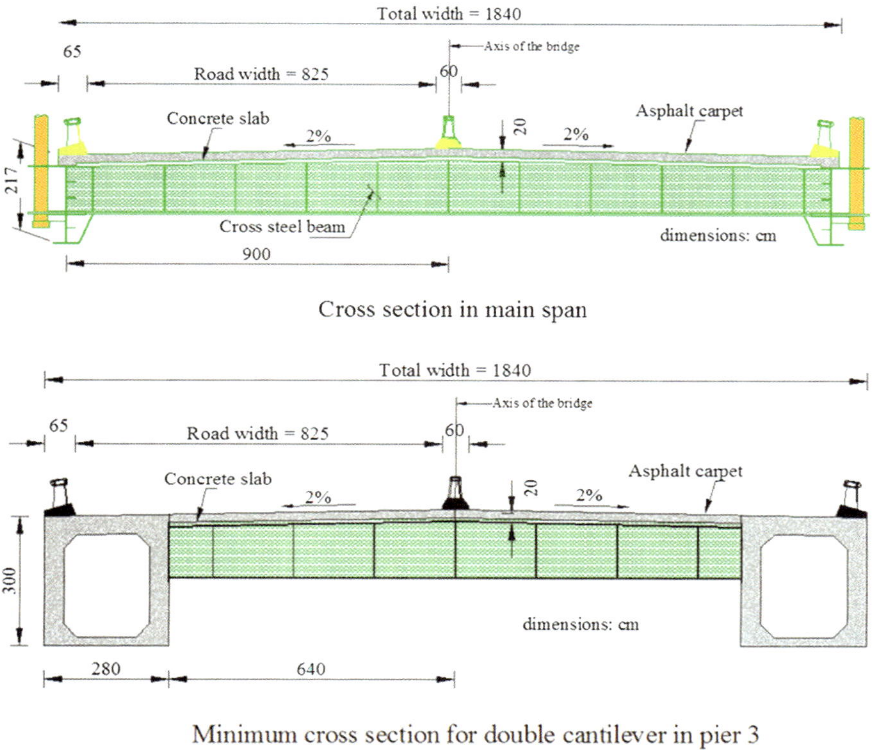

Fig. 4 Cross sections of the superstructure, "Carrizo" bridge

3 Load Testing

After the construction of the bridges, static, dynamic and impact tests were carried out [5, 6]. In this paper we only present results of tests under static loads produced by trucks placed at different locations on the main spans. The aim was to produce strains in key locations of the superstructures; six and nine tests were carried out, each one with a different load configuration, on the "Baluarte" and "Carrizo" bridges, respectively. Five and three axle trucks were used during the tests (see Fig. 5).

Fiber optic sensors were installed to record strains. For the "Baluarte" bridge 22 sensors were used and distributed along the main span distributed in three sections: Sinaloa, Center and Durango; for the "Carrizo" bridge 16 sensors, 8 on each cable stay span. Also, additional temperature sensors were placed on both bridges. Economic constraints limited the number of sensors. Using topographic devices, deflections were measured simultaneously during the static tests on both bridges.

For purposes of showing the type of results obtained during the tests, below we present some results of experimental results of static load tests. For each test,

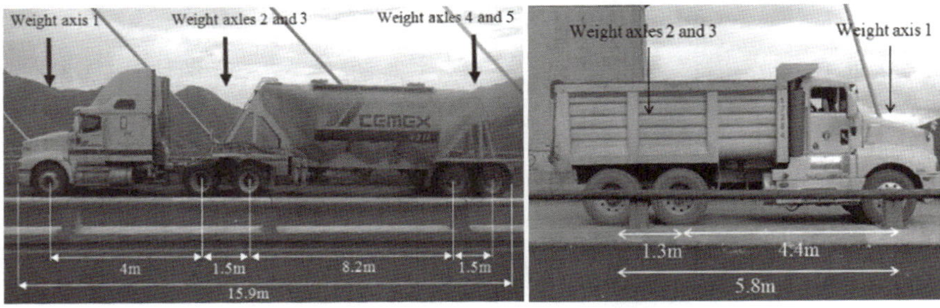

Fig. 5 Type of trucks used for field tests

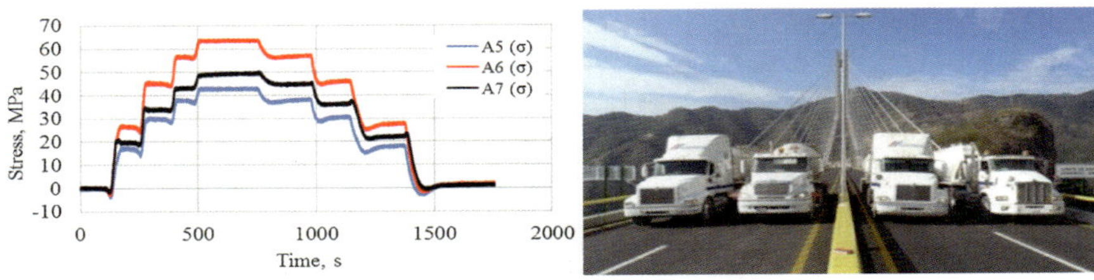

Fig. 6 Static load test "E2", "Baluarte" bridge

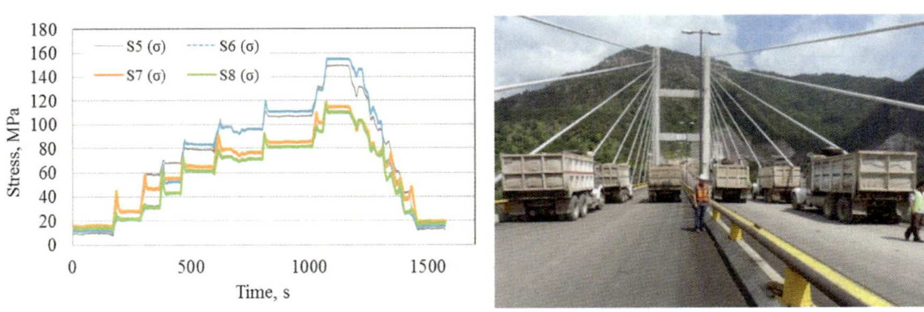

Fig. 7 Static load test "E9", "Carrizo" bridge

the trucks configuration is presented together with a graph showing the obtained stresses considering a structural steel A572 gr50. These graphs only show the results for the sensors at the center of the main spans. The Fig. 6 shows the configuration of trucks on the deck and the corresponding measurements of a specific test performed at the "Baluarte" bridge; sensors A5, A6 and A7 were placed at the center of the mains span. Figure 7 shows another test carried out at the "Carrizo" bridge; in this case sensors S5, S6, S7 and S8 were placed at segments twelve and thirteen, side Sinaloa. All of these sensors were located at the bottom flange of the girders.

In addition to the static tests, dynamic tests were also performed with trucks passing at different speeds.

4 Mathematical and Experimental Results

It is well known that cable stayed bridges are complex structures for modeling, even though modern tools of computing are available. Finite element model of the main girders and piers were developed using standard beam elements and plate finite elements for the concrete slab [7]. To account for the geometric nonlinear behavior due to cable sag effects, cables were modeled with special elements [8]; also, before the analysis of the static and dynamic tests, it was necessary to perform a non-linear static analysis [9] to compute the tangent stiffness of the bridge in its dead-load deformed state. Piers and towers were fixed at the base. The same load conditions of static and dynamic field tests (E1, E2, etc.), were simulated with the mathematical models and strains and deflection were determined for each of them.

Ambient vibration tests were also carried out to estimate the modal shapes and frequencies; this modal information was also used to design the aeroelastic model of the complete bridge, to identify modes susceptible to wind loading, and to provide information about vibration frequencies and modal shapes for the wind tunnel tests [3, 4].

4.1 Comparison of Results

The structural assessment of the superstructures was achieved by comparing experimental and analytical data obtained from the developed mathematical model. Tables 1 and 2 show some results of actual and simulated stresses for the "Baluarte" and "Carrizo" bridges, respectively. For Table 1, all results are associated to the same sensor. It is observed that maximum differences are around 13 %. For the Carrizo bridge results, this difference is higher. However both comparisons show a reliable matching of the comparison.

Topographic monitoring of vertical displacements was also carried out. Table 3 shows some results registered at the "Baluarte" bridge. Results for the "Carrizo" are not presented because in some cases they were affected by the sensibility of

Table 1 Experimental and mathematical results of stresses. "Baluarte" bridge

Test	Model MPa	Instrumentation MPa	Difference	
			MPa	%
E1	23.3	21.9	1.4	6.1
E2	78.8	68.0	10.8	13.7
E5	103.6	89.8	13.8	13.4

Table 2 Experimental and mathematical results of stresses. "Carrizo" bridge

Test	Model	Instrumentation	Difference	
	MPa	MPa	MPa	%
E1	32.7	26.6	6.1	18.7
E5	−8.2	−8.8	0.6	−7.5
E9	155.0	154.0	1.0	0.6

Table 3 "Baluarte" bridge, deflections, static tests

Test	Load condition	Math. model	Experimental	Difference	
		mm	mm	mm	%
E1	2T[a] in parallel	111	59	41	47
E2	4T in parallel	215	206	−6	4
E5	4T in pairs	208	190	10	9

[a]*T* truck

the measuring device. Results of experimental and mathematical frequencies associated to different components of vibration were also compared. For the vertical direction of vibration the comparison was very reliable, for both bridges.

4.2 Analysis of the Response Under Moving Loads

During field testing at the "Carrizo" bridge, the passage of trucks imposed significant vibrations of the deck. Using the calibrated mathematical model of the bridge, a direct integration time history analysis was performed and the dynamic effects were evaluated. A nine axle design truck was used as the moving load, running at different velocities. The results of the analysis are presented in Fig. 8 for two directions of movement of the vehicle: Durango-Sinaloa and Sinaloa-Durango, and at selected points on opposites sides of the bridge deck, both at about 132 m from the central pylon. Point B is located near the double cantilever section of the bridge superstructure; point A is located on the opposite side. Time history records of accelerations were also produced

"Peaks" shown in Fig. 8 are associated to the passage of vehicles at the point of interest (A or B) and the effect that causes on the other side of the deck (points A or B); at one side the deck moves downward and the other upward. It is also observed that the maximum values of displacement, for both directions of circulation, are close to 15 cm. With respect to accelerations, the maximum values were about 40 cm/s^2.

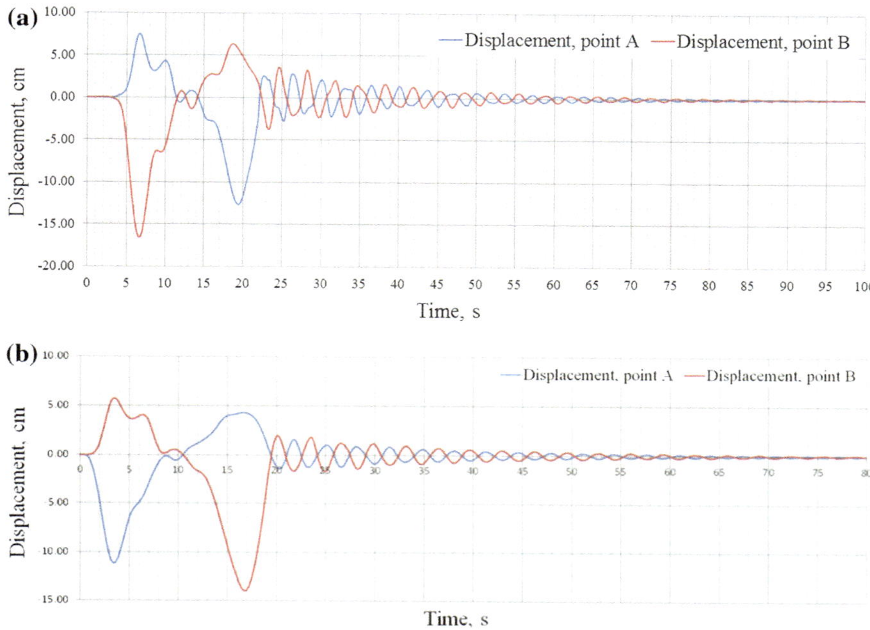

Fig. 8 Time histories of displacements, "Carrizo" bridge. **a** Direction Sinaloa. **b** Direction Durango

5 Conclusions

Modelling, analysis and experimental testing of two cable stayed bridges were presented. Measured deflections at the "Baluarte" bridge are comparable with the results of the mathematical model. Maximum stresses produced by live loads are about 90 MPa, and when added to dead load stresses the results are below design or allowable stresses. For the "Carrizo" bridge, stresses produced by live loads, similar to design loads, are higher, about 156 MPa, which represents 74.2 % of the allowable stress. In this case, when this level of stresses is considered together with dead loads, some sections of the structure fail to comply, by a minimum amount, with allowable design stresses.

Although both superstructures showed a very good elastic recovery capacity, from the point of view of displacements and vibrations the "Carrizo" bridge superstructure is more flexible compared to the "Baluarte" bridge. In general, both superstructures showed a good performance.

The calibrated mathematical models and the instrumentation will be used for a permanent monitoring of the behavior of the superstructures under regular traffic.

Acknowledgments The Ministry of Communication and Transportation of Mexico provided the funds for the development of the work carried out by IIUNAM staff.

References

1. Ministry of Communication and Transport of Mexico (2008) Executive project of "Baluarte" bridge. Mexico
2. Ministry of Communication and Transport of Mexico (2011). Executive project of "Carrizo" bridge. Mexico
3. Gómez R, Pozos A (2010) Report on the wind tunnel testing at UWO of an aeroelastic model of the "Baluarte" bridge (in Spanish). Technical report presented to the Ministry of Communication and Transport of Mexico. Institute of Engineering, UNAM, Mexico
4. Gómez R, Pozos A (2012) Report on the wind tunnel testing at UWO of an aeroelastic model of the "Carrizo" bridge (in Spanish). Technical report, presented to the Ministry of Communication and Transport of Mexico. Institute of Engineering, UNAM, Mexico
5. Gómez R, Arenas LM, Sanchez R, Escobar JA, Paredes CA, Rosales ON (2013) Static and dynamic load testing of the main span of the "Baluarte" cable stayed bridge (in Spanish). Technical report, presented to the Ministry of Communication and Transport of Mexico. Institute of Engineering, UNAM, Mexico
6. Gómez R, Arenas LM, Sanchez R, Escobar JA, Paredes CA, Rosales ON (2013) Static and dynamic load testing of the main span of the "Carrizo" cable stayed bridge (in Spanish). Technical report presented to the Ministry of Communication and Transport of Mexico, Institute of Engineering, UNAM, Mexico
7. SAP2000 (2009) V. 14.0. Integrated finite element analysis and design of structures. Computers and structures, Berkeley, CA
8. Gimsing NJ (1983) Cable supported bridges, concept and design. Wiley, New York
9. Nazmy AS, Abdel-Ghaffar AM (1992) Effects of ground motion spatial variability on the response of cable-stayed bridges. Earthq Eng Struct Dynam 21:1–20

Deriving a Load Model for the Braking Force on Road Bridges: Comparison Between a Deterministic and a Probabilistic Approach

Joao Martins, M.-A. Fénart, G. Feltrin, A.-G. Dumont and K. Beyer

Abstract The current provisions for the braking force in the Eurocodes are much more demanding than previously enforced national codes from most European countries. Hence, safety assessments of existing bridges may show a lack of compliance with the new safety requirements of current maintenance codes, making the relevant authorities responsible for their strengthening. The current load model for the braking force was derived from a deterministic evaluation of traffic configurations with characteristics that do not correspond with actual traffic measurements. This procedure differs from models describing vertical traffic loads, where the effects of traffic were analysed within a probabilistic framework in compliance with the partial safety factor design method. This paper investigates the advantages, limitations and hypotheses of a probabilistic model for the braking force when compared with deterministic models. Original results for the characteristic value of the braking force on road bridges are presented, which consider a realistic time-history of bridge crossing vehicles generated by a traffic microsimulation tool with input data from a Swiss Weigh-In-Motion station. The results are compared with the braking force of the Eurocodes, highlighting the role of the probability of braking and the influence of the dynamic characteristics of bridges on the characteristic value of the braking force.

J. Martins (✉) · G. Feltrin
Empa, Swiss Federal Laboratories for Materials Science and Technology, Dübendorf 8600, Switzerland
e-mail: joao.martins@empa.ch

M.-A. Fénart · A.-G. Dumont
Traffic Facilities Laboratory, EPFL, Lausanne 1015, Switzerland

K. Beyer
Earthquake Engineering and Structural Dynamics Laboratory, EPFL, Lausanne 1015, Switzerland

A. Caner et al. (eds.), *Developments in International Bridge Engineering*, Springer Tracts on Transportation and Traffic 9,
DOI 10.1007/978-3-319-19785-2_3

1 Introduction

Braking forces are horizontal loads due to traffic that have to be considered in the design of new road bridges and in the assessment of existing ones. The magnitude of these forces can be computed according to the load models provided by design codes. In recent codes, such as the Eurocodes and the American code "AASHTO LRFD Bridge Design Specifications", safety verifications are based on the assessment of limit states by the partial safety factor design method, which, in the American nomenclature, is termed load and resistance factor design method (LRFD). In this framework, a structure is deemed reliable if limit states are not exceeded in any selected design situations, the computation of which includes the design values of the basic variables, namely actions, material properties and geometrical data.

In the partial safety factor design method, the characteristic value is the principal representative value of an action. For variable actions, it is multiplied by partial factors to compute the other representative values—combination, frequent and quasi-permanent values. This yields loading levels with different probabilities of exceedance, which are grouped into "design scenarios". The design scenarios and corresponding partial safety factors are determined in design codes, e.g. in "Eurocode—Basis of structural design" (EC0) [1]. In Merzenich and Sedlacek's [2] background studies of "Eurocode 1: Actions on structures—Part 2: Traffic loads on bridges" (EC1-2) [3], the characteristic values of vertical traffic loads were calibrated for a 1000-year return period.

The partial safety factor method of checking the structural safety of structures for a certain design scenario has, therefore, a semi-probabilistic nature. On the one hand, it is implied that the computation of the representative values of the actions involved a probabilistic analysis. On the other hand, its application is deterministic, in the sense that it is assumed that the structure is reliable if the safety assessments of the relevant limit states are checked, and not by evaluating its reliability index. However, probabilistic analyses were not used to derive the representative values of all types of actions: According to Annex C of EC0 [1], deterministic methods have primarily been used in the first generation of the Eurocodes when relevant statistical data was lacking. For these actions, the load models were derived on the basis of historical events and subjective assumptions regarding extreme scenarios, instead of following a probabilistic approach.

Regarding traffic loads on bridges, the derivation of vertical load models 1 and 2 for global and local verifications, respectively, was based on a probabilistic analysis of traffic data from a Weigh-in-Motion station in Auxerre, France. These load models fit thus into the framework of the partial safety factor design. The braking force, however, was derived from deterministically defined traffic scenarios. Since the model lacks a sound probabilistic background, the return period of the braking force defined in the code is unclear. This means that, unlike the case of vertical traffic loads, there is no measure of the probability of exceedance of the braking force model, which, therefore, might be either non-conservative, hence, unsafe, or too conservative, hence, uneconomical.

This paper describes different approaches to the derivation of braking force models on road bridges. Section 2 describes three procedures underlying models enforced by European design codes. All these models are based on deterministically defined traffic situations. Presenting a new probabilistic approach, Sect. 3 describes the structure of a stochastic model developed to compute the characteristic value of the braking force considering the randomness of traffic situations and of driver behaviour. The implementation of this model is illustrated using results from a traffic flow simulation. The paper concludes with a discussion of the key parameters of the probabilistic model and an outlook on future work.

2 Derivation of Braking Force Models in Design Codes—A Deterministic Approach

Before partial safety factors were adopted in design codes, the effects of deterministically defined traffic situations were used to derive traffic load models on bridges. This applied in particular to the load models for the braking force that were adopted by the 1970 revision of the Swiss code SIA 160 and by the British Standard BS 5400 (1978). The derivation of these models is presented in Sects. 2.1 and 2.2, respectively. Also for EC1-2 [3], despite the nonconformity with the vertical load models, a deterministic approach was applied to derive the braking force model, which is described in Sect. 2.3.

2.1 Swiss Standards SIA 160 (1970)

In 1964, in order to derive a load model for the braking force that would be part of the revised Swiss standard SIA 160 (1970), a total of 26 experiments were performed with different initial speeds, number of vehicles and distances between vehicles [4]. The time-history of the deceleration of each vehicle was measured using tachographs, accelerometers and photographic sequences. The magnitude of the maximum braking force transmitted to the pavement $F_{\mathrm{b\,max}}$ was estimated by Newton's laws of motion:

$$F_{\mathrm{b\,max}} = m \cdot \max_{t} \left\{ \sum_{i=1}^{N} a_i(t) \right\} \quad (1)$$

where m is the mass of the vehicles, which was assumed equal for all vehicles, N is the number of vehicles, and $a_i(t)$ is the deceleration over time t of the *ith* vehicle. Figure 1 shows the results of applying Eq. 1 using the measured data from convoys with initial speeds of 30 and 60 km/h and $m = 15$ t. Figure 1 also shows that the maximum braking force of the load model adopted by the revision

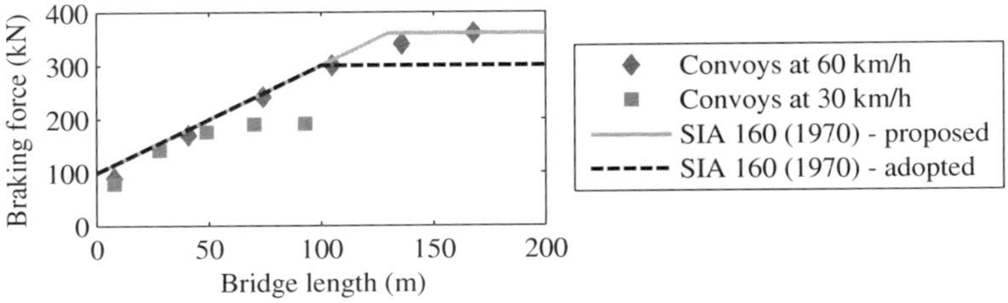

Fig. 1 The braking forces in the load model from SIA 160 (1970) and in deterministically defined braking situations. Adapted from [4]

commission differs from the one proposed in the study. Although the maximum braking force in the proposed model was 360 kN, which was attained when six vehicles travelling with an initial speed of 60 km/h braked simultaneously, the revision commission adopted a maximum of 300 kN, which corresponds to a convoy of four vehicles instead.

2.2 British Standards BS 5400 (1978)—HA Model

In 1969, Burt [5] developed a new braking force model for British Standard BS 5400 (1978), which replaced BS 153 (1954). This model became one of the most severe in European codes, and for a significant range of bridge lengths it is even more severe than the one proposed in EC1-2 [3].

For small bridge lengths, the model reflects the braking action of single axles carrying as much as 18 tonnes of load and of individual vehicles weighing between 16 and 32 tonnes. The braking forces due to convoys of up to four vehicles braking were also computed with a vehicle weight of 25 tonnes and an initial speed of 45 mph (72 km/h). The driver reaction time was taken equal to 0.7 s. Two different values were used for the deceleration of the vehicles, 0.5 g and 0.75 g, where g stands for the acceleration due to gravity. Results of the braking forces due to one or more vehicles braking is represented in Fig. 2 as $m_{Convoy} \cdot 0.5$ g and

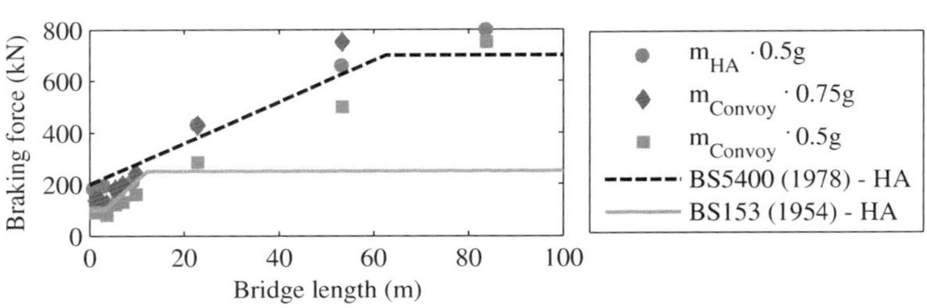

Fig. 2 The braking force in the British standards BS 153 (1954) and BS 5400 (1978), and in deterministically defined braking situations. Adapted from [5]

$m_{Convoy} \cdot 0.75$ g, for decelerations of 0.5 and 0.75 g respectively. In the same figure, results for $m_{HA} \cdot 0.5$ g represent the application of a deceleration of 0.5 g to the mass that corresponds to the force computed according to the HA vertical load model from the British Standards. Figure 2 also shows the previously existing and the afterwards implemented load models for the braking force.

2.3 Eurocode 1—Part 2 (2003)

The derivation procedure of the braking force model in EC1-2 [3] is described by Sedlacek et al. [6]. As in the previous cases, the model reflects the braking of deterministically defined traffic configurations. Up to five vehicles are assumed to travel at the same speed of 81 km/h and to achieve the same deceleration of 0.5 g. The adopted driver reaction time is 1.0 s and the maximum assumed lorry weight is 40 tonnes.

The dynamic response of the bridge to braking forces was modelled with elastic, viscously damped single-degree-of-freedom (SDOF) systems, with a range of natural vibration periods from 0.1 s to 5.0 s and a damping ratio of 0.07. The excitation of the SDOF system is the result of the superposition of impact loads corresponding to the braking of each vehicle. The systems were analysed using closed form solutions for elastic SDOFs and the maximum displacement of each system was determined. Based on these results, dynamic amplification factors of 1.8, 1.4 and 1.2 were taken, respectively, for one, two, and three or more vehicles in convoy. In this way, the derivation of the load model does not depend on the dynamic properties of a particular bridge, yet it considers dynamic effects as a function of the number of vehicles in convoy. Figure 3 shows the computed braking forces for convoys of one to five vehicles with 25 tonnes (m = 25 t), 30 tonnes (m = 30 t) and 40 tonnes (m = 40 t) simultaneously braking as well as the EC1-2 [3] load model. Figure 3 illustrates that the load model roughly envelopes, for increasing bridge lengths, the braking events with one 40-tonne vehicle, two and three 30-tonne vehicles, four and five 25-tonne vehicles and, for bridges longer than 200 m, five 30-tonne vehicles.

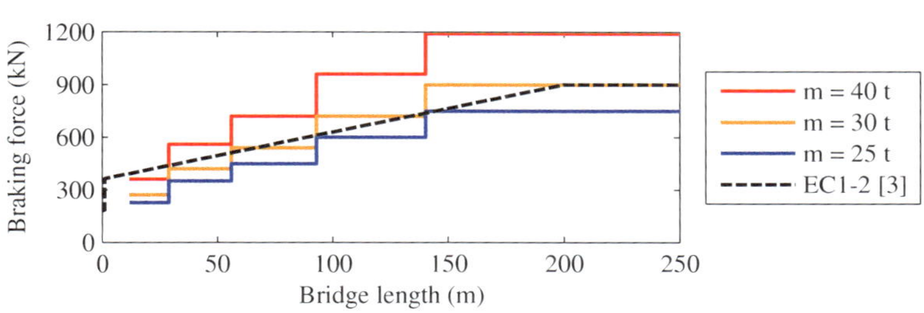

Fig. 3 The braking force in EC1-2 [3] and in deterministically defined braking situations. Adapted from [6]

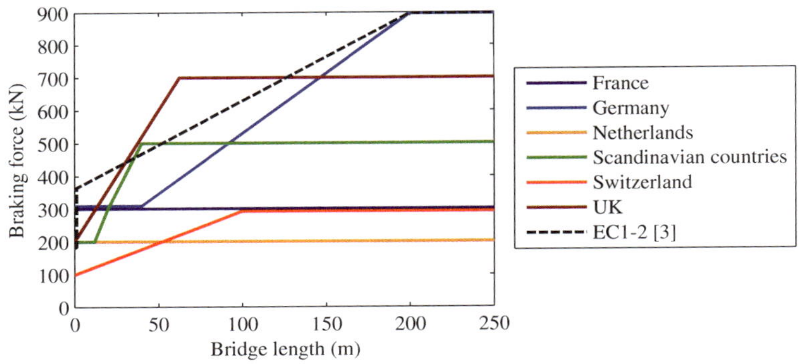

Fig. 4 The braking force in EC1-2 [3] and in other European codes. Adapted from [2]

In contrast with the derivation procedures described in the previous sections, the dynamic response of the bridge was considered when defining the braking force model of EC1-2 [3]. It is also the load model that accounts for the heaviest lorries, by considering 40-tonne vehicles. This explains why EC1-2 [3] braking force model is more severe than any of the previously enforced in European countries, at least for some ranges of bridge length, if not for all [2].

With the lightest vehicles (15 t) and only static computations, the Swiss code is understandably the least severe of the three given examples. Moreover, the braking forces in the background studies of the Swiss code were computed with experimental measurements of decelerations over time, whereas the other load models reflect the conservative assumption of having all vehicles simultaneously braking with their maximum deceleration. Furthermore, the British Standard stands out for medium bridges because much shorter inter-vehicle distances were considered than in the other codes. Figure 4 compares the EC1-2 [3] model with these standards and some other national design codes previous to the Eurocode.

3 Probabilistically Derived Braking Force Model

The use of probabilistic tools in modern codes minimizes the role of subjective assumptions in the derivation of load models. Croce and Sanpaolesi [7] explain this direction of code development further by outlining the derivation of the vertical traffic load models included in EC1-2 [3]. These models were calibrated so that their target effects on the structure have a quantified probability of exceedance [7]. In the Eurocodes, the vertical traffic load models are calibrated to produce effects with a 1000-year return period, i.e., 5 and 10 % probability of being exceeded in 50 and 100 years, respectively.

In the literature, multiple studies using probabilistic techniques can be found, both to derive or validate load models for codes (e.g., [2, 7–9]), and to assess the reliability of existing structures (e.g., [10–13]). However, in the scope of traffic loads, these studies focus only on the effects of vertical forces on bridges and not on horizontal forces. Even in deterministic analyses, the effects of vehicles braking on bridges have seldom been addressed, and when they have, for instance in [14–16], the increase in the vertical effects due to the redistribution of axle loads is the main object of analysis, not the horizontal effects.

The next section presents a probabilistic model that realistically takes into account the characteristics of traffic flow instead of relying on deterministically chosen traffic configurations. Driver behaviour, vehicle response to severe braking and horizontal dynamic response of bridges to braking vehicles are also considered. Moreover, the probability of having a braking event on a bridge must be included in order to compute the quantile of the braking force distribution that corresponds to its characteristic value. The purpose of this procedure is to compute the value of the braking force that corresponds to the same return period as the vertical traffic load models.

3.1 Structure of the Probabilistic Model

A time-history of bridge crossing vehicles, resulting from microsimulation of traffic, is analysed and used to generate one-lane traffic configurations. These traffic configurations are, in fact, vehicle convoys where each vehicle's speed, length and weight is known as well as the inter-vehicle distances, and the correlations between these variables are respected.

Each traffic configuration is used to model at least one braking event, which initiates with the first vehicle in convoy applying full brakes until it stops. Values for reaction time and maximum deceleration of the following vehicles are randomly generated, within the limits of driver response and of the braking systems' properties, so that the vehicles brake to a full halt avoiding collision. A collision would involve mechanisms of dissipation of energy other than friction between tyres and pavement, hence presumable smaller braking loads. The probability distribution of driver reaction times is based on Olson's observations [17], and bilinear deceleration profiles are applied, with a deceleration rise time of 0.25 and 0.6 s combined with maximum decelerations up to 9 and 7 m/s^2, for light and heavy vehicles, respectively. The differences in the deceleration profiles are due to the weight of the vehicles as well as to the nature of the braking systems employed in light and heavy vehicles. While light vehicles use hydraulic brakes, heavy vehicles employ pneumatic brakes, which lead to longer deceleration rise times [18].

As in similar studies, such as the Eurocodes' background calculations [2] or O'Brien and Enright's [8] determination of the aggressiveness of traffic, it is assumed that the statistical properties of traffic remain unchanged in time. Since driver behaviour and vehicle response are also taken as time-invariant, the braking processes have a stationary nature. Therefore the probabilistic framework used for vertical load models also applies for braking forces. Furthermore, all traffic configurations, independently of the position on the bridge, are considered equally prone to generate a braking event. This is not an essential condition; however, differentiating probabilities of braking along the length of the bridge would require a thorough analysis of the triggering events behind braking situations. Such an analysis would need to be based on data from on-road traffic studies, which monitor driver behaviour and vehicle performance in their natural environment and may include accelerometers, video logging and GPS locators. Since this is a relatively new field of research, published results are scarce and the required data has not been made available yet. For an overview of the potentialities of these studies, the reader is referred to Carsten et al. [19].

As in Sedlacek et al. [6], the bridge was reduced to a linear elastic, viscously damped SDOF system, which enables the use of closed-form expressions to compute its dynamic response. The excitation force transmitted to the deck of a bridge in case of a braking event depends on the traffic scenario, i.e., a traffic configuration on an initial position. It would be inefficient to build the probability distribution of the braking force considering just the number of braking events that are likely to happen during the period of time of the microsimulation. Long periods would be needed and most traffic configurations would not be used. On the other hand, the statistical properties of all traffic characteristics are preserved stipulating equally spaced initial positions and using the same set for all traffic configurations. Since it is assumed that a braking event is equally probable for all the traffic scenarios thus defined, if all of them are used in the stochastic model, the statistical properties of the probability distribution of the braking force are also respected. The more initial positions are considered for initiating braking events, the more values of the braking force are computed, which enriches the probability distribution but takes considerable computational effort and, for sufficiently close initial positions, may have no further influence in the computation of the characteristic value.

Having computed the probability distribution of the braking force, the characteristic value of the braking force corresponds to the quantile of this distribution associated to a return period of 1000 years. To compute this quantile, the probability of occurrence of the events modelled has to estimated, i.e. the probability that a driver, while covering a distance equal to the spacing between the modelled initial positions of braking events, initiates a braking event by applying full brakes until the vehicle comes to a halt, forcing following drivers to brake too. In this study, this concept was designated as braking probability and identified by the letter β. Given this definition, for N traffic scenarios the expected number of braking events is therefore simply $\beta \cdot N$. This number is also the expected number of braking events during the period τ that corresponds to the period covered by the microsimulation of traffic. The expected number n of braking events in 1000 years, which is a period of time $1000/\tau$ times larger than the period of the microsimulation for τ in years, is given by:

$$n = \frac{1000 \cdot \beta \cdot N}{\tau} \tag{2}$$

Finally, the characteristic value of the braking force, which according to the definition of return period is the magnitude of the force that is exceeded on average once in 1000 years, is given by the quantile of the distribution of the braking force that has a probability of non-exceedance P:

$$P = 1 - \frac{1}{n} \tag{3}$$

3.2 Results Using Traffic Records

Traffic records from a Weigh-in-Motion (WIM) station located on the A2 Swiss motorway, near Monte Ceneri, between Lugano and Bellinzona, from the year 2008, were combined with records from a close by Swiss Automatic Road Traffic Count (SARTC) station, from the month of June of the same year. These two sources of traffic data were used to generate a realistic time-history of bridge crossing vehicles using the traffic simulation software AIMSUN [20]. The data set from a WIM station is not enough on its own since only heavy vehicles are recorded. Whereas the effect of light vehicles' weight is negligible by comparison with that of heavy vehicles, they play a relevant role in the way the sequence of braking actions occurs in a convoy of vehicles. On the other hand, SARTC records use a very broad classification for vehicles and do not record their weights.

The microsimulation tool generated 163,182 vehicles, representing one week of traffic, 63,181 on the left lane and 100,001 on the right lane. All vehicles are assigned a speed and headway with respect to the previous vehicle and are identified by a class number. The six classes of light vehicles are the same as in the "Swiss 10" classification [21], which is also used by the SARTC stations. The same classification uses only four categories for heavy vehicles. Instead 14 categories were used in the microsimulation, computed according to the spacing between axles and vehicle weights, similarly to the work of Meystre and Hirt [22]. The greater detail in the classification of heavy vehicles assures the relative proportions between different classes, increases the accuracy of the random generation of vehicle length and weight, which is obtained from bootstrapping WIM records of each class, and eliminates spurious records that do not comply with any of the predefined classes.

Figure 5 compares the braking force model from EC1-2 [3] with the characteristic values of the braking force computed for a fundamental period of the bridge in the longitudinal direction T_b of 5.0 s, a damping ratio of 0.07, bridge lengths ranging from 25 to 250 m and β between 10^{-9} and 10^{-5}. As initial positions of braking events were taken 10 m apart, these values for the braking probability denote a rate for the initiation of braking events, applying full brakes until the vehicle stops, of once in 10^7 and 10^3 km, respectively.

These results, which are specific for the Swiss motorway A2 near Monte Ceneri, show characteristic values of the braking force as high as 1050 kN, significantly more severe than the 900 kN of the load model of EC1-2 [3], for bridges of a length larger than 150 m and for a braking probability of 10^{-5}. Overall, for short bridges, the load model of EC1-2 [3] leads to similar values as the probabilistic model with a braking probability of 10^{-8}, whereas for long bridges a braking probability of almost 10^{-6} reproduces more adequately the loading level of the code. These values of β correspond to initiating a braking event applying full brakes on average once in 1,000,000 and 10,000 km travelled, respectively. Also, if β is smaller than 10^{-8}, all the values of the braking force given by the current load model have a return period larger than 1000 years. In that case, the enforced braking force model is conservative and, hence, uneconomical in comparison with the vertical load models. The opposite applies for β larger than 10^{-6}, for which the braking force model of EC1-2 [3] is clearly below the safety target used to derive the vertical traffic load models.

To assess the influence of the dynamic characteristics of the bridge on the characteristic value of the braking force, Fig. 6 shows the results computed for 100 m long bridges with fundamental periods in the longitudinal direction T_b between 0.1 and 5.0 s. It shows that for more flexible bridges, which is the case in Fig. 5, the dynamic response is clearly more severe.

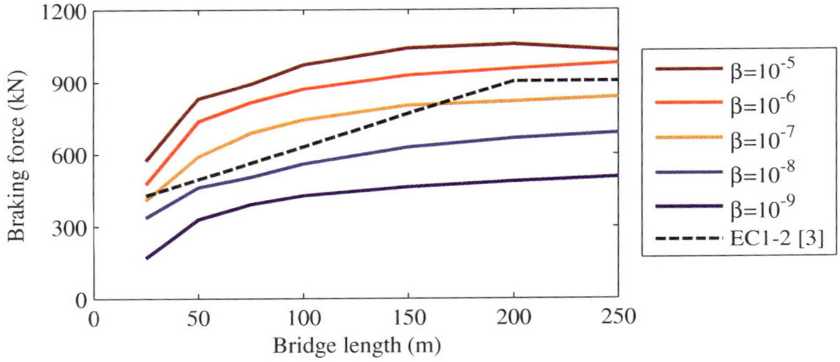

Fig. 5 The braking force in EC1-2 [3] and results from stochastic simulations

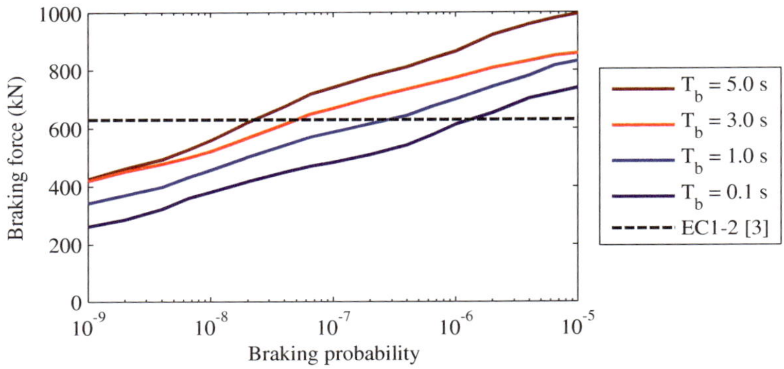

Fig. 6 Characteristic values of the braking force on a 100 m long bridge, for different natural vibration periods and braking probabilities. Comparison with EC1-2 [3]

Both figures illustrate that the results of the probabilistic model are very sensitive to the assumed braking probability. The calibration of the braking probability is therefore of the utmost importance, as well as validating the assumption that it is independent of vehicle class and traffic configuration or, alternatively, introducing different braking probabilities in a way that best reproduces driver behaviour.

4 Conclusions

Deterministic approaches to the derivation of load models are simple to apply once the target scenarios have been defined. However, the distinction between an assumed worst-case scenario and a worst-case scenario likely to happen in the lifetime of the structure is highly subjective. The examples in this paper show how similar premises in the definition of braking events, convoys of vehicles with the same weight and maximum deceleration, and drivers with the same reaction time, yield very different results. In new codes a different approach based on sound probabilistic considerations should be adopted to derive load models.

In fact, to comply with the probabilistic framework of modern codes, thus having all safety verifications harmonised for the same reliability level, the derivation of load models must take into account the variability of the design actions and be able to compute high percentiles of the probability distribution of their effects, namely their characteristic value. This value, pivotal to the design of structures, corresponds to rare events, with a small but quantified probability of occurrence during their lifetime.

In the scope of load derivation, traffic loads are particularly challenging because modelling them combines several random variables that are correlated, which can be accounted for with microsimulation of traffic. When computing braking forces, additional sources of randomness include driver reaction time, maximum deceleration applied by following vehicles and the probability of having a braking event. Their quantification and realistic inclusion in the presented probabilistic model represents one of the main challenges of this research. In the end, the leverage of the probabilistic approach, when compared with the deterministic one, is to be data-oriented, which is lost if subjective judgements play decisive roles in the modelling assumptions.

Given the lack of data, the use of deterministic approaches has always been preferred to derive load models for the braking force. However, not only traffic measurements amount now to several years and microsimulation of traffic has been successfully applied in structural verifications, but the number of on-road traffic studies is increasing. Therefore, a probabilistic approach, such as the model presented in this paper, is now feasible to compute the characteristic value of the braking force on bridges. The proposed model uses realistic braking processes and traffic configurations based on measurements from a Swiss motorway near Monte Ceneri to compute a probability distribution of the braking force. Since the computation of the braking probability is still ongoing, results are presented as a function of this parameter and compared to the load model present in the Eurocodes. Hence, results are not conclusive with regard to its actual return period and, as a consequence, safety level.

In a next stage, considering traffic from different sites with significant flow of heavy vehicles and a representative variety of structural schemes, a final load model can be derived, fulfilling the Eurocodes' objectives in terms of safety assessment independently of bridge type and location. Moreover, this probabilistic approach can also be applied for reliability assessment of existing bridges, adequately adjusting the return period and giving as input a site-specific time-history of crossing vehicles computed with local traffic records.

Acknowledgments The research work was financed through a grant by the Federal Roads Office in Switzerland to project AGB 2011-03 "Updated braking forces for the assessment of road bridges". This contribution is gratefully acknowledged.

References

1. EN 1990:2002+A1 (English version): Eurocode—basis of structural design. CEN, European Committee for Standardization, Brussels (2002)
2. Merzenich G, Sedlacek G (1995) Hintergrundbericht zum Eurocode 1—Teil 3.2: "Verkehrslasten auf Straßenbrücken". Bundesminister für Verkehr, Abt. Straßenbau: Bonn-Bad Godesberg, Germany
3. EN 1991-2:2003 (English version): Eurocode 1: actions on structures—part 2: traffic loads on bridges. CEN, European Committee for Standardization, Brussels (2003)
4. Rapport sur les essais de freinage des 2 et 3 juin 1964 [effectués par le] Laboratoire de statique des constructions de l'EPUL Lausanne [pour la] Commission de révision des normes 160 [de la] Société suisse des ingénieurs et des architects. EPUL (École Polytechnique Universitaire de Lausanne), Lausanne (1965)
5. Burt ME (1969) RRL Technical note no. TN 401—forces on bridges due to braking vehicles. Design Division—Road Research Laboratory, Crowthorne, Berkshire
6. Sedlacek G, Merzenich G, Paschen M, Bruls A, Sanpaolesi L, Croce P, Calgaro JA, Pratt M, Jacob, LM, Boer, VA, Hanswille G (2008) Background document to EN 1991—part 2—traffic loads for road bridges—and consequences for the design. JRC—European Commission
7. Croce P, Sanpaolesi L (2005) Chapter 1: bridges—actions and load combinations in handbook 4: design of bridges—guide to basis of bridge design related to Eurocodes supplemented by practical examples. Available at http://eurocodes.jrc.ec.europa.eu/
8. O'Brien EJ, Enright B (2013) Using weigh-in-motion data to determine aggressiveness of traffic for bridge loading. J Bridge Eng 18(3):232–239
9. Nowak AS (1995) Calibration of LRFD bridge code. J Struct Eng 121(8):1245–1251
10. Caprani CC (2012) Calibration of a congestion load model for highway bridges using traffic microsimulation. Struct Eng Int 22(3):342–348
11. Crespo-Minguillón C, Casas JR (1997) A comprehensive traffic load model for bridge safety checking. Struct Saf 19(4):339–359
12. Bailey S (1996) Basic principles and load models for the structural safety evaluation of existing road bridges. Ph.D. thesis no. 1467. EPFL (École Polytechnique Fédérale de Lausanne), Lausanne
13. Červenka V (2013) Reliability-based non-linear analysis according to fib model code 2010. Struct Concr 14(1):19–28

14. Gupta RK, Traill-Nash RW (1980) Bridge dynamic loading due to road surface irregularities and braking of vehicle. Earthq Eng Struct Dynam 8(2):83–96
15. Yang YB, Wu YS (2001) A versatile element for analyzing vehicle–bridge interaction response. Eng Struct 23(5):452–469
16. Azimi H, Galal K, Pekau OA (2013) A numerical element for vehicle-bridge interaction analysis of vehicles experiencing sudden deceleration. Eng Struct 49:792–805
17. Olson PL, Cleveland DE, Fancher PS, Schneider LW (1984) Parameters affecting stopping sight distance. University of Michigan Transportation Research Institute, Washington, DC
18. Limpert R (2011) Brake design and safety. SAE International, Warrendale
19. Carsten O, Kircher K, Jamson S (2013) Vehicle-based studies of driving in the real world: the hard truth? Accid Anal Prev 58:162–174
20. TSS Transportation Simulation System. Aimsun version 6.1 user's manual. TSS Transportation Simulation System, Barcelona (2011)
21. ASTRA (Federal Roads Office) (2009) Verkehrszähler—Richtlinie. Available at www.astra.admin.ch
22. Meystre T, Hirt MA (2006) Evaluation de ponts routiers existants avec un modèle de charge de trafic actualisé. Mandat de recherche AGB 2002/005. Federal Roads Office, Lausanne

The Revisions for the Gusset Plate Load Rating Approach: A Comparative Case Study

Burak Boyaci and Burcu Guldur

Abstract In 2009, the Federal Highway Administration (FHWA) published the "Load Rating Guidance and Examples for Bolted and Riveted Gusset Plates in Truss Bridges". According to this document, main truss member gusset plates were required to be examined for five limit states based on either the Load Factor Rating (LFR) or the Load and Resistance Factor Rating (LRFR) approaches. However, further investigations showed that this methodology generally resulted in conservative gusset plate ratings. The National Cooperative Highway Research Program (NCHRP) was sponsored to address these concerns by investigating the shear, tensile and compressive strength resistance of gusset plates at the strength limit state. The findings of this program resulted in a set of propositions for revisions to the American Association of State Highway and Transportation Officials (AASHTO) Manual for Bridge Evaluation to be implemented in 2014. It is observed that these new provisions for the LFR and the LRFR methods result in significant variation in rating results for the different limit states. In this paper, a comparative study is completed on a gusset plate from the Charles M. Braga Jr. Memorial Bridge, which is located in Fall River/Somerset, MA, U.S.A., to investigate the dissimilarities between the load rating results obtained based on both the 2009 and the 2014 provisions for the LFR approach. The primary aim of this work is to provide a better understanding on the proposed changes for the gusset plate load rating approach described in the Manual for Bridge Evaluation, 2nd Edition, 2014 Interim Revisions.

B. Boyaci (✉)
TranSystems Corporation, 38 Chauncy St Suite 200, Boston, MA 02111, USA
e-mail: bboyaci@transystems.com

B. Guldur
Department of Civil Engineering, Hacettepe University, Ankara 06800, Turkey

A. Caner et al. (eds.), *Developments in International Bridge Engineering*,
Springer Tracts on Transportation and Traffic 9,
DOI 10.1007/978-3-319-19785-2_4

1 Introduction

The I-35 W highway bridge over the Mississippi River in Minneapolis, MN collapsed on August 1, 2007. Subsequently, the National Transportation Safety Board (NTSB) initiated an investigation to identify the likely initiation cause for the failure and its location. A FHWA report prepared by Ocel and Wright [1], which was an outcome of this investigation, showed that several of the gusset plates had been over stressed. Particularly, the inelastic buckling of one of these plates was found to be the probable cause of the failure. After this failure, evaluating the safety of the gusset plates on this type of bridges became an urgent priority. Thus, the FHWA released the "Load Rating Guidance and Examples for Bolted and Riveted Gusset Plates in Truss Bridges (2009 FHWA Guidance)" in 2009 that provides guidelines to State transportation departments [2]. The 2009 FHWA Guidance includes sections, which are in accordance with both the LRFR method and the LFR method.

The 2009 FHWA Guidance was based on the existing provisions in the AASHTO LRFD Bridge Design Specifications [3] and the older AASHTO Standard Specifications for Highway Bridges [4]. The method given in this guidance was proved to yield in conservative gusset plate ratings. To investigate this, the FHWA initiated a research project conducted by the NCHRP. The results of this research are collected under the "Guidelines for the Load and Resistance Factor Design and Rating of Riveted and Bolted Gusset-Plate Connections for Steel Bridges" [5] that presents the findings of an experimental and analytical investigation analyzing the failure modes of steel truss bridge gusset plated connections.

The project conducted by the NCHRP, NCHRP Project 12–84, was primarily focused on the development of an LRFR approach to the gusset plate rating. The translation of the resistance factors stated by the NCHRP Project 12–84 to the LFR philosophy is not trivial since the live load models are different. However, the findings of the NCHRP Project 12–84 still resulted in some changes in the LFR approach for gusset plate ratings compared to the approach discussed in the 2009 FHWA Guidance. These changes are given in the "Manual for Bridge Evaluation, 2nd Edition, 2014 Interim Revisions (2014 MBE Interim)" [6]. In this paper, a comparative study, which presents the gusset plate rating of a specific gusset plate from Braga Bridge in accordance with the LFR approach for both the 2009 FHWA Guidance and the 2014 MBE Interim, is given.

2 Charles M. Braga Jr. Memorial Bridge

The gusset plate investigated in this paper is a part of Charles M. Braga Jr. Memorial Bridge. This bridge extends between the town of Somerset and the city of Fall River, near the mouth of the Quequechan River. Figure 1 shows the

Fig. 1 The location of the bridge on the map, and an image of the north elevation of the bridge [7]

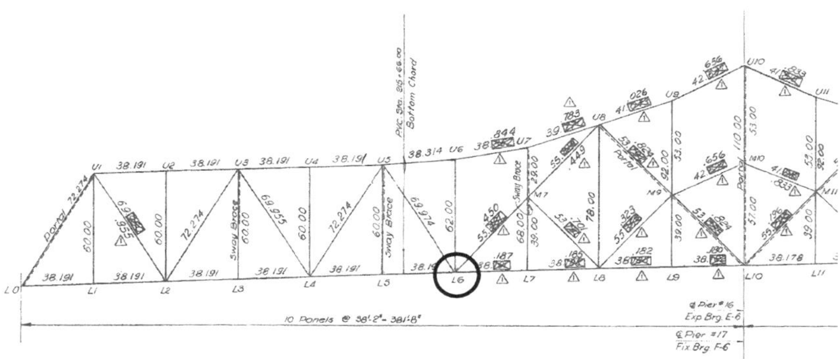

Fig. 2 Drawing of the Braga Bridge where the investigated gusset plate L6 is shown

location of the bridge on the map and an image of the north elevation of the bridge. The gusset plate labeled as L6, which is shown in Fig. 2, is used in this study.

The Braga Bridge has been under a rehabilitation program, which is funded by the Massachusetts Department of Transportation, since 2003. The successive load rating analyses of the gusset plates have been performed by TranSystems Corporation since 2010.

3 Comparative Results

In this work, the gusset plate shown in Fig. 2 is examined for five limit states based on the LFR approach. These limit states are completed by applying necessary checks for shear, tension, compression, fasteners and splice. For the investigated gusset plate, the calculations are carried on for both the 2009 and the 2014 provisions for the LFR approach. Here, the 2009 provisions refer to the LFR approach of the 2009 FHWA Guidance and the 2014 provisions refer to the LFR approach of the 2014 MBE Interim. The differences between the obtained results for both provisions are individually discussed in the following sections.

The load rating equations of the LFR approach are given by Eqs. 1 and 2, [8]. The capacity of the gusset plates indicated by R_r is reduced by 10 % in the 2009 provisions for increasing the safety margin.

$$RF_{INV} = (R_r - 1.3 \cdot P_{DL}) \div 2.17 \cdot P_{LL} \tag{1}$$

$$RF_{OPR} = (R_r - 1.3 \cdot P_{DL}) \div 1.3 \cdot P_{LL} \tag{2}$$

where

RF_{INV} = Inventory load rating factor,
RF_{OPR} = Operating load rating factor,
R_r = Capacity,
P_{DL} = Dead load,
P_{LL} = Live load imposed by HS20 rating truck.

4 Shear Resistance

The first limit state of the gusset plate load rating calculation is the shear capacity check. For the shear capacity check, there are several changes in the 2014 provisions for the LFR approach in comparison with the 2009 provisions.

The first change is in the Ω factor used for the shear yielding check. The equation for calculating the shear yielding capacity is given in Eq. 3. The Ω factor is either 0.74 or 1.00 in 2009 provisions. $\Omega = 0.74$ is used for the case of flexural stress distribution whereas $\Omega = 1$ is used for the case of uniform shear distribution. For the investigated gusset plate, conservatively, Ω is taken as 0.74 assuming that the stress distribution is flexural. In the 2014 provisions, the Ω factor is defined as a single value which is 0.88. Therefore, since $\Omega = 0.74$ is used originally, with the updated provision the rating for shear yielding increases.

$$R = 0.58 \cdot F_y \cdot A_g \cdot \Omega \tag{3}$$

where

F_y = Yield strength of the plate (ksi),
A_g = Gross area (in.^{2}),
Ω = Shear reduction factor.

The second difference between the 2009 and 2014 provisions is the change in the definitions for the shear planes, where the shear failures are expected to occur. The full length of cross-sections, which are shown in Fig. 3 by AA_{NET}, BB_{NET} and

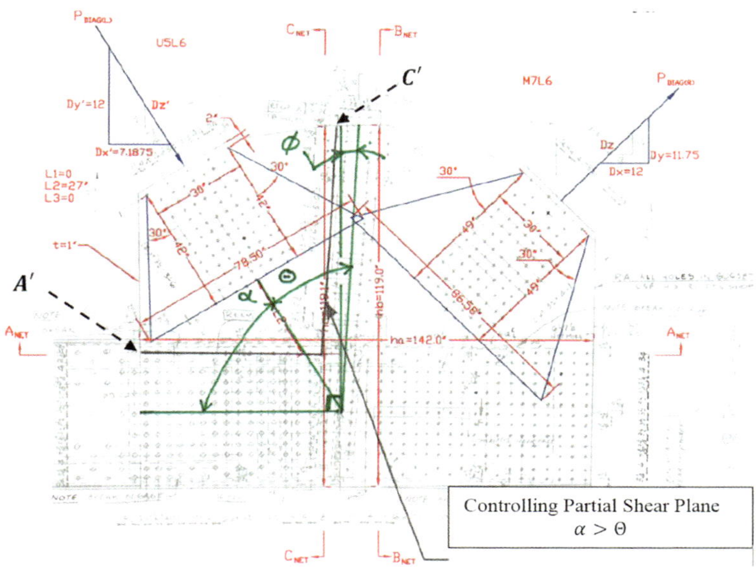

Fig. 3 Drawing of the L6 gusset plate where shear planes are represented

CC_{NET}, are used for shear resistance calculations in the 2009 provisions. However, the partial shear planes AA' and CC' are the cross-sections used for shear resistance calculations in the 2014 provisions. These partial shear planes are shown in Fig. 3. The lengths of AA' and CC', which are taken along the member fastener lines, are defined as the distances between plate edges and other fasteners lines. The partial shear planes are only checked around compression members. For shear resistance computation, only the controlling partial shear plane is used. Since there is no chamfered end of the compression member, the governing criteria for controlling partial shear plane is the angle between the compression member and the adjoining members [6].

The final dissimilarity between the two discussed provisions is that the 2014 provisions exclude the checks for the shear rupture for the partial shear planes. The 2014 provisions stated that the shear rupture check is not required for the partial shear planes.

The load rating results for shear resistance obtained by using both 2009 and 2014 provisions are given in Table 1. As shown in this table, the *AA* and *BB* shear planes are not checked in the 2014 provisions. The only check performed is on the CC' shear plane that represents the controlling partial shear plane. Since the *AA* and *BB* shear plane checks are eliminated from 2014 provisions, the overall rating for shear resistance results being less conservative compared to the 2009 provisions. However, when only the load ratings obtained as results of the *CC* and CC' shear plane checks are compared, the 2014 provisions yield into more conservative load ratings.

Table 1 The load rating results for shear resistance both for 2009 and 2014 provisions

Shear plane	Shear yielding capacity, kips (kN)	Shear rupture capacity, kips (kN)	Rating factor	
			Inventory	Operating
2009 provisions				
AA	5607.1 (24,941.6)	7597.1 (33,793.5)	0.77	1.28
BB	4698.9 (20,901.7)	6300.7 (28,026.9)	2.16	3.61
CC	4702.8 (20,919.1)	6307.3 (28,056.2)	2.22	3.71
2014 provisions				
CC′	3542.3 (15,756.9)	Not applicable	1.46	2.43

5 Tensile Resistance

Tension check consists of two subsections; the first check is for the yield capacity, and the second one is for the block shear. There is no change for the yield capacity check between the 2009 and the 2014 provisions. However, the equation that is used to calculate the block shear capacity is updated. Equations 4 and 5 are the block shear resistance equations given in the 2009 provisions. Whereas, the Eq. 6 displays the updated block shear resistance equation given in the 2014 provisions.

The former equations rely on an initial criteria based on the ratio of the net tension and the net shear areas. On the other hand, the recent provisions provide a generalized equation independent of the ratio of the net tension and shear areas.

$$If A_{tn} \geq 0.58 A_{vn}, \quad then\, R_r = 0.85(0.58\, F_y A_{vg} + F_u A_{tn}) \tag{4}$$

$$If A_{tn} < 0.58 A_{vn}, \quad then\, R_r = 0.85(0.58\, F_u A_{vn} + F_y A_{tg}) \tag{5}$$

$$R_r = \min\left(\phi_{bs} R_p (0.58 F_u A_{vn} + F_u A_{tn}), \phi_{bs} R_p \left(0.58 F_u A_{vg} + F_u A_{tn}\right)\right) \tag{6}$$

where

$\phi_{bs} = 0.85$, resistance factor for block shear rupture,
$R_p =$ Reduction factor for holes taken equal to 0.90 for bolt holes punched full size and 1.0 for bolt holes drilled full size or subpunched and reamed to size,
$A_{vg} =$ Gross area along the plane resisting shear stress (in.^{2}),
$A_{tg} =$ Gross area along the plane resisting tension stress (in.^{2}),
$A_{vn} =$ Net area along the plane resisting shear stress (in.^{2}),
$A_{tn} =$ Net area along the plane resisting tension stress (in.^{2}),
$F_u =$ Tensile strength of the plate (ksi).

The load rating results for tensile resistance obtained by using both the 2009 and the 2014 provisions are given in Table 2. As shown in this table, there is no change in the yielding capacity results. However, the capacity increases for the right diagonal member while the block shear capacity of the vertical member decreases using the equation given in the 2014 provisions. Therefore, the effect of the revisions is case dependent, and the rating results cannot be estimated.

Table 2 The load rating results for tensile resistance both for 2009 and 2014 provisions

Member	Yielding capacity, kips (kN)	Block shear rupture capacity, kips (kN)	Rating factor	
			Inventory	Operating
2009 provisions				
Vertical member	7852.2 (34,928.3)	8030.8 (35,722.7)	29.73	49.62
Right diagonal	7965.4 (35,431.8)	7337.8 (32,640.1)	2.65	4.42
2014 provisions				
Vertical member	7852.2 (34,928.3)	7203.8 (32,044.1)	30.34	50.64
Right diagonal	7965.4 (35,431.8)	7466.7 (33,213.5)	3.40	5.68

6 Compressive Resistance

The third limit state of the gusset plate load rating calculation is the compressive resistance check. In order to calculate the compressive resistance, plate stability is checked at the gusset plate zones in the vicinity of compression members.

A Whitmore section is used to determine the effective width of the idealized compression member. The unbraced length of the idealized compression member is defined as the standoff distance of the compression member relative to the surrounding members. This distance is defined as the average of $L1$, $L2$ and $L3$ shown in Fig. 4a in the 2009 provisions. However, in the 2014 provisions, it is given as the distance from the middle of the Whitmore section to the nearest member fastener line in the direction of the member L_{mid}. L_{mid} is shown in Fig. 4b.

In addition, the effective length factor K values in the 2009 provisions were selected based on the expected buckled shape of the idealized compression

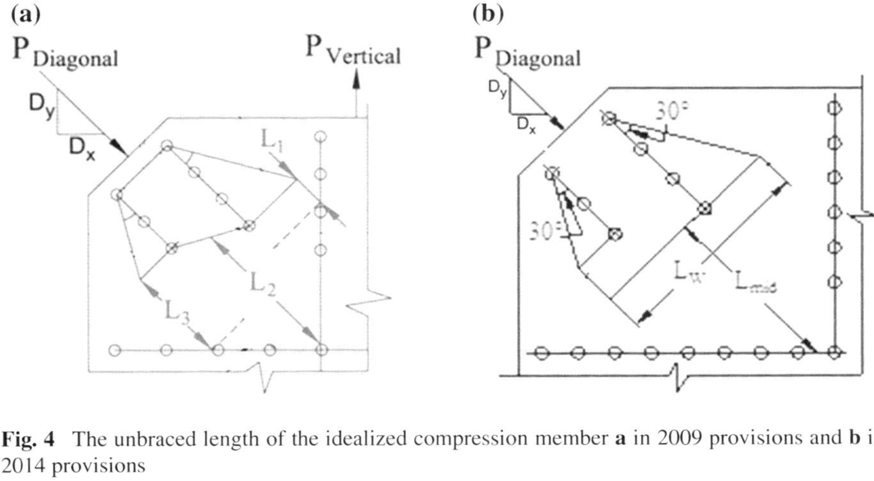

Fig. 4 The unbraced length of the idealized compression member **a** in 2009 provisions and **b** in 2014 provisions

Table 3 The load rating results for compressive resistance both for 2009 and 2014 provisions

Member	Compression capacity, kips (kN)	Rating factor	
		Inventory	Operating
2009 provisions			
Left diagonal	5898.8 (26,239.1)	2.63	4.39
Chord	6423.7 (28,574.1)	5.16	8.61
2014 provisions			
Left diagonal	5599.1 (24,906.1)	2.94	4.91

member. The design K value can vary in between 0.65 and 2.0. However, in the 2014 provisions, the K value is given as a constant value for all cases ($K = 0.5$).

The load rating results for the compressive resistance obtained by using both the 2009 and the 2014 provisions are given in Table 3. In the 2014 provisions, for which a more stable gusset plate with a higher compression capacity is assumed, the effective length factor K value is always smaller than the K values given in the 2009 provisions. However, the unbraced length definition in the 2014 provisions results in a longer length than the one defined in the 2009 provisions, resulting in a decrease in the compression capacity. As shown in Table 3, the effect of the change in the unbraced length definition is more dominant than the effect of the change in the K value definition, therefore the compression capacity decreases. Although the compression capacity decreases for the left diagonal member, the rating factor increases due to the 10 % capacity reduction in the load rating equations given in the 2009 provisions. The compression capacity check of the chord splices is discussed under a new section in the 2014 provisions, which is later in this paper.

7 Fasteners

The fourth limit state of the gusset plate load rating calculation is the fastener resistance check. Although there are no updates on the procedures to be followed for the fastener resistance check in the 2014 provisions; the shear strength of rivets was revised in 2011 MBE Interim [9]. As shown in Table 4, while the bearing capacity of the rivets remains unchanged, their shear capacity slightly increases due to a revision in the shear strength of the rivets.

8 Chord Splices

The last limit state needs to be checked for the gusset plate load rating calculations is the chord splice resistance check. Chord splices are checked under tensile or compressive resistance limit states in the 2009 revisions. However, a new

Table 4 The load rating results for fastener resistance both for 2009 and 2014 provisions

Member	Fastener shear capacity, kips (kN)	Fastener bearing capacity, kips (kN)	Rating factor	
			Inventory	Operating
2009 provisions				
Vertical member	2638.9 (11,738.4)	13,507.2 (60,083.1)	8.82	14.72
Left diagonal	5042.3 (22,429.2)	25,808.4 (114,801.4)	1.80	3.00
Right diagonal	5277.9 (23,477.2)	27,014.4 (120,166.1)	1.03	1.72
2014 provisions				
Vertical member	2814.9 (12,521.2)	13,507.2 (60,083.1)	10.78	17.99
Left diagonal	5709.5 (25,397.1)	25,808.4 (114,801.5)	3.06	5.11
Right diagonal	5711.5 (25,406.1)	27,014.4 (120,166.1)	1.87	3.12

Table 5 The load rating results for chord splice resistance for 2014 provisions

Member	Compression capacity, kips (kN)	Rating factor	
		Inventory	Operating
2014 provisions			
Chord	6423.7 (28,574.1)	5.16	8.61

procedure is introduced in the 2014 provisions to better represent the capacities of the chord splices. In the new procedure, the spliced section consisting of gusset plate and the splice plates are assumed to resist the combined stress, which is a result of the applied axial load and bending due to the eccentricity caused by the offset of the resultant force from the centroid of the section [6]. See Table 5 for the chord splice capacities and the corresponding rating results calculated according to the 2014 provisions.

9 Conclusions

In this paper, the load rating results are presented for five limit states individually based on the LFR approach. The results for these five limit states are presented for both 2009 and 2014 provisions. Later, the 2009 and the 2014 provisions are compared and the changes between these two provisions are shown. The reflections of the changes on the results are discussed. It has been shown that these changes may affect the obtained end results of the gusset plate rating. Thus, it is important to differentiate these changes and their effect on the results in order to properly assess the obtained load rating results.

For future work, the comparison of the LFR and the LRFR approaches for gusset plate load rating is planned to be discussed. The same procedure will be followed to display the results on a specific gusset plate. This will provide a complimentary discussion on the available gusset plate evaluavtion approaches.

Acknowledgments Funding for this research is provided by TranSystems Corporation. Any opinions, findings, and conclusions expressed in this material are those of the authors and do not necessarily reflect the views of the sponsor. Authors would also like to thank Massachusetts Department of Transportation for giving permission to reprint the calculations from the Braga Bridge Project.

References

1. Ocel J, Wright W (2008) Finite element modeling of I-35 W bridge collapse. Interim Draft Rep Turner-Fairbank Highw Res Cent Rep Federal Highw Adm
2. Federal Highway Administration (FHWA) (2009) Load rating guidance and examples for bolted and riveted gusset plates in truss bridges
3. American Association of State Highway and Transportation Officials (AASHTO) (2007) LRFD bridge design specifications, 4th edn
4. American Association of State Highway and Transportation Officials (AASHTO) (2002) Standard specifications for highway bridges, 17th edn
5. National Transportation Safety Board (NTSB) (2013) Guidelines for the load and resistance factor design and rating of riveted and bolted gusset-plate connections for steel bridges
6. American Association of State Highway and Transportation Officials (AASHTO) (2014) Manual for bridge evaluation, 2nd edn, 2014 interim revisions
7. Wikipedia (2014) Charles M. Braga Jr. memorial bridge
8. American Association of State Highway and Transportation Officials (AASHTO) (2011) The manual for bridge evaluation, 2nd edn
9. American Association of State Highway and Transportation Officials (AASHTO) (2011) Manual for bridge evaluation, 2nd edn, 2011 interim revisions

The Use of Appropriate Finite Element Modelling Techniques for the Idealization of a Novel Hybrid Multi-girder Bridge Deck System

Donna Chen and Raafat El-Hacha

Abstract Examination into FEA techniques for modelling the performance of a specially designed hybrid multi-girder bridge deck system with three model configurations of varying complexities showed good correlation with experimental global load-deflection behavior as well as local strain behavior at the base of the bridge system. For ascertaining peak load conditions, the simplest of the three model configurations was found to provide the best accuracy of results where the peak condition was estimated within a 2–12 % accuracy range for the FEMs studied. However, the local behavior in the transverse composite system, particularly in the horizontal GFRP rods, obtained from FEA greatly underestimated (3.5–9 times) the tensile strain experienced, even when tube-in-tube elements were implemented to represent the unbonded movement of GFRP rod through the surrounding plastic ducts. Good representation of overall behavior in the transverse composite system, however, was obtained with tube-to-tube elements, with transverse curvature exhibited. Interestingly, the addition of tube-to-tube elements did not affect the overall global performance exhibited through FEA, resulting in identical load-deflection curves with a comparable FEM without the use of tube-to-tube elements. The computational time required to complete the model with tube-to-tube elements was also less than that required for the model without tube-to-tube elements. Further study is recommended to determine and eliminate the unknown constraint placed on the GFRP rods, to allow for greater accuracy in the prediction of tensile strain.

D. Chen · R. El-Hacha (✉)
University of Calgary, 2500 University Drive N.W, Calgary T2N1N4, Canada
e-mail: relhacha@ucalgar.ca

D. Chen
e-mail: dsmchen@ucalgary.ca

A. Caner et al. (eds.), *Developments in International Bridge Engineering*,
Springer Tracts on Transportation and Traffic 9,
DOI 10.1007/978-3-319-19785-2_5

1 Introduction

The characterization of composite materials, and their application in structural members, is a rapidly growing field of research. Fibre Reinforced Polymer (FRP) materials have gained popularity in structural applications, firstly as internal reinforcement in reinforced concrete members [1], then for the purposes of strengthening and rehabilitating existing structural elements [2]. Advancements in technology and state-of-the-art knowledge allowed for further progress, with the use of full structural shapes (such as plates and beams) composed of FRP material acting as the primary load carrying structural member. Many bridge designs using FRP materials, including hybrid structures, have also been proposed and tested at various international research institutions [3]; the aim of these designs is to achieve an optimized (structurally and financially) structural member by incorporating the most desirable and advantageous qualities of each material component in the design.

This research study is focused on the performance of a novel steel-free multi-girder hybrid bridge deck system, composed solely of FRP materials and Ultra-High Performance Concrete (UHPC). Due to the resources and complexities required for large-scale experimental testing, it is not feasible to conduct comprehensive testing as a means to evaluate the performance of this new proposed design. Finite element analysis (FEA) techniques can be applied to idealize the behavior of the hybrid bridge deck design through commercial software packages such as ABAQUS. The development of a FE model (FEM), including model validation with experimental results obtained from testing of a quarter-scale prototype specimen, will be discussed in this paper. In an effort to optimize the required computation power with the accuracy of FEA calculations, specific attention will be placed on the influence of model simplifications on obtained results, with comparisons made between FEMs of increasing complexity.

2 Description of Hybrid Bridge Deck System Design

The main objective of this design was to achieve a steel-free, and thus corrosion-resistant, hybrid bridge deck system composed solely of high performance materials. The feasibility of combining individual hybrid beams [4], made up of FRP components as well as UHPC, into a three-dimensional (3D) hybrid bridge deck system was investigated with this unique multi-girder design. Four parallel Glass FRP (GFRP) hollow box beams were spaced evenly in the transverse direction, each reinforced on the tensile face with a sheet of adhesively bonded Carbon FRP (CFRP). An integrating UHPC slab was cast-in-place overtop of the multi-girder system. Previous experimental work proved the effectiveness of a specially designed bond interface between the top flange of the GFRP and the bottom surface of the UHPC slab. This bond system consists of coarse silica sand aggregates bonded directly to the coarsened surface of the GFRP top flange prior to casting of UHPC, which enhanced bond strength and performance at the GFRP-UHPC interface when compared with the use of only epoxy adhesive bonding [4]. Premature debonding failure was prevented when the specially designed bond system was implemented. Particular interest was devoted to the design of the transverse composite action system, required to effectively distribute loads between the parallel girders as well as to allow the various structural components to perform as a unified system. The completed design, with dimensions in mm, is provided in Fig. 1.

The uniqueness of this type of hybrid bridge deck system design does not have a readily available set of design guidelines and standards. It is important to note that one special feature of UHPC is that, by including internal steel fibres, it removes the required use of internal steel reinforcement bars while possessing a tensile strength more than five times that of normal reinforced concrete material. Design standards for multi-girder bridge systems found in the Canadian Highway Bridge Code (CSA S6-06) [6], intended for the design of bridge system constructed using conventional material such as steel reinforced concrete and structural steel sections, specify requirements for horizontal straps directly below the concrete slab in cases where internal reinforcement is not provided. There are also additional requirements for the placement of transverse diaphragms at set intervals along the length of the system. To further enhance the transverse performance of the hybrid bridge deck system design, the use of horizontal unbonded post-tensioned (PT) GFRP rods (Fig. 1), through the use of plastic ducts around the GFRP rods, to provide an initial camber was also examined. In order to satisfy the original requirement of a steel-free design, the transverse composite action system was achieved using GFRP threaded rods [7]. Interpretations of the strength and serviceability requirements in CSA S6-06 [6] were made for the first design iteration of the transverse composite system in the hybrid bridge deck system.

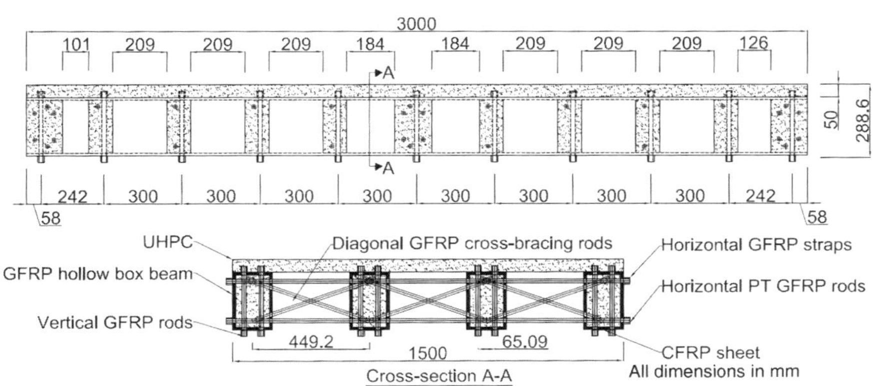

Fig. 1 Detailed design and dimensioning of hybrid bridge deck system [5]

3 Experimental Testing and Results

The hybrid bridge deck system was tested using four simultaneous wheel loads positioned concentrically about the longitudinal and transverse axes, at a center-to-center spacing of 600 and 450 mm, respectively. A detailed examination of the experimental test results was discussed in another publication [5]. A brief summary of relevant results will be presented here. Under ultimate loading conditions, the hybrid bridge deck system reached an initial peak load of 590 kN at an actuator displacement of 32 mm after which progressive failure occurred. Symptoms of local failures during progressive failure include: buckling of the GFRP cross-bracing rods, web shearing of GFRP hollow box beams, tensile cracking in the CFRP and GFRP at the bottom face of the interior girders and tensile cracking on the top surface of the UHPC slab. A maximum displacement of 60 mm was reached with sustained load resistance provided by the bridge deck system when testing was stopped to prevent damage to the instrumentation positioned below the specimen.

4 Development of Finite Element Model

The 3D FE model was developed using ABAQUS 6.10. Due to symmetry of the hybrid bridge deck system dimensions and the loading conditions, only half of the specimen was modelled. The GFRP shapes and CFRP sheets were modelled as continuum shell elements with the GFRP rods (vertical, horizontal and diagonal) modelled using linear beam elements. The remaining components (UHPC and steel plates) were modelled as linear solid elements. A diagram showing the completed model, with the mesh density used, is provided in Fig. 2.

A detailed summary of the material properties used is listed in Tables 1, 2, 3 and 4. To reflect the influence of damage on the major structural components, Hashin damage (GFRP) and concrete damaged plasticity (UHPC) properties were also implemented.

5 Model Investigation Using FEA Techniques

Deflection controlled loading was applied in the FEMs by administering a uniform displacement boundary condition at the steel loading plates. Quasi-static load steps were used to provide numerical convergence past the peak load and through the zone of progressive failure. For this preliminary study, the post-tensioning strain, equal to an average value of 1100 $\mu\varepsilon$ [5], was not applied. A variety of model configurations was implemented in the FEM, each with their own individual set of assumptions and simplifications. Due to the size of the model, it

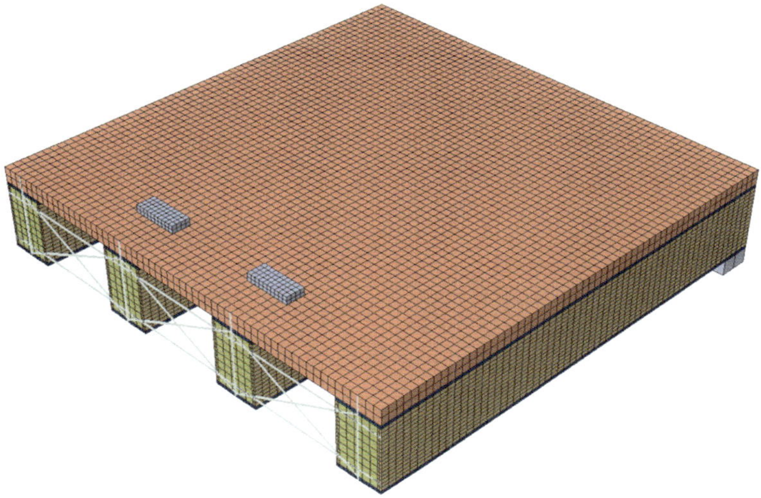

Fig. 2 Completed 3D FE model for hybrid bridge deck system

Table 1 Material property parameters for UHPC slab

Concrete elasticity	
Modulus of elasticity: E (MPa)	50,000
Poisson's ratio: ν	0.2
Concrete compression hardening [8]	
Compressive stress (MPa)	Plastic strain (−)
96.5	0.0000000
110.3	0.0000284
137.9	0.0000720
165.5	0.0001410
189.6	0.0004140
Concrete tension stiffening [8]	
Maximum tensile stress (MPa)	Plastic strain (+)
15.9	0.00000
15.9	0.00836
0.0	0.00900
Concrete damage plasticity parameters [8]	
Dilation angle	15°
Eccentricity	0.1
Ratio of initial equibiaxial to initial uniaxial compressive yield stress: *fb0/fc0*	1.16
Ratio of the second stress invariant on the tensile meridian: K	2/3
Viscosity parameter	0

Table 2 Material property parameters for GFRP hollow box beam [4]

GFRP elasticity	
Longitudinal modulus of elasticity: E_1 (MPa)	21,000 (flanges), 16,000 (webs)
Transverse modulus of elasticity: E_2 (MPa)	5520
In-plane shear modulus: G_{12} (MPa)	2930
Through-thickness longitudinal shear modulus: G_{13} (MPa)	2930
Through-thickness transverse shear modulus: G_{23} (MPa)	2930
Poisson's ratio: ν	0.33
GFRP Hashin damage	
Longitudinal ultimate stress: σ_1 (MPa)	207 (tension), 207 (compression)
Transverse ultimate stress: σ_2 (MPa)	48.3 (tension), 103 (compression)
In-plane ultimate shear stress: τ_{12} (MPa)	31
Through-thickness ultimate shear stress: τ_{23} (MPa)	31

Table 3 Material property parameters for CFRP sheet, GFRP rods and steel plates

Parameters	CFRP [8]	GFRP rods [7]	Steel plates
Modulus of elasticity: E (MPa)	95,800	21,000	200,000
Poisson's ratio: ν	0.3	0.29	0.2

Table 4 Material property parameters for GFRP-UHPC bond interface [9]

Bond interface elasticity/stiffness	
Mode I material stiffness: K_{nn} (MPa/m)	200
Mode II material stiffness: K_{ss} (MPa/m)	600
Mode III material stiffness: K_{tt} (MPa/m)	600
Quadratic stress traction-separation parameters	
Mode I nominal stress: σ_n (MPa)	2
Mode II nominal stress: σ_s (MPa)	6
Mode III nominal stress: σ_t (MPa)	6
Linear energy damage evolution power law	
Mode I fracture energy: G_I (kJ/m)	0.3
Mode II fracture energy: G_2 (kJ/m)	0.9
Mode III fracture energy: G_3 (kJ/m)	0.9

is important to ascertain the degree of complexity required to achieve a validated model with an acceptable level of accuracy.

5.1 Model Configuration #1—No UHPC Stiffener Blocks

This model evaluated the FEA results obtained when internal UHPC stiffener blocks were not implemented. At the diaphragms, tie constraints were applied between the end nodes of each GFRP cross-bracing rod and the corresponding interior top or bottom surface of the GFRP hollow box beam flange. The horizontal GFRP straps (top) and PT GFRP rods (bottom) were tied at the end nodes with the outermost surfaces of the outer GFRP girders. The vertical GFRP rods were similarly tied to the top surface of the UHPC slab and bottom face of the CFRP sheet.

5.2 Model Configuration #2—With UHPC Stiffener Blocks

This second model iteration included the presence of internal UHPC blocks. To more accurately idealize the bond fixity between the GFRP cross-bracing rods and the UHPC stiffener blocks, embedded constraints were imposed on the interior nodes of the GFRP cross-bracing rods. A hard contact (no penetration) interaction constraint was placed between the outer surface of the interior UHPC stiffener blocks and the interior surfaces of the GFRP hollow box beam. Similarly, to provide partial fixity between the GFRP hollow box beam and the UHPC stiffener block, a penalty friction constraint was imposed with a friction coefficient of 0.9999. Through previous studies, this friction constraint has been found to provide sound results [10].

5.3 Model Configuration #3—With Tube-to-Tube Elements for Horizontal GFRP Rods

An increased level of complexity was added to this model, to restrict the movement of the horizontal GFRP straps and PT GFRP rods. Tube-to-tube elements were added to these horizontal components with the slide line prescribed through the GFRP hollow box beam webs and the UHPC stiffener blocks. A clearance of 1.0 mm all around the horizontal GFRP rods was also imposed to simulate the condition of a rod moving freely through the plastic duct.

6 Analysis and Discussion of FEA Results

6.1 Comparison of Global Performance

Comparison between the global load-deflection behavior of the system obtained experimentally and through FEA, with the deflection at the steel plates shown, is presented in Fig. 3.

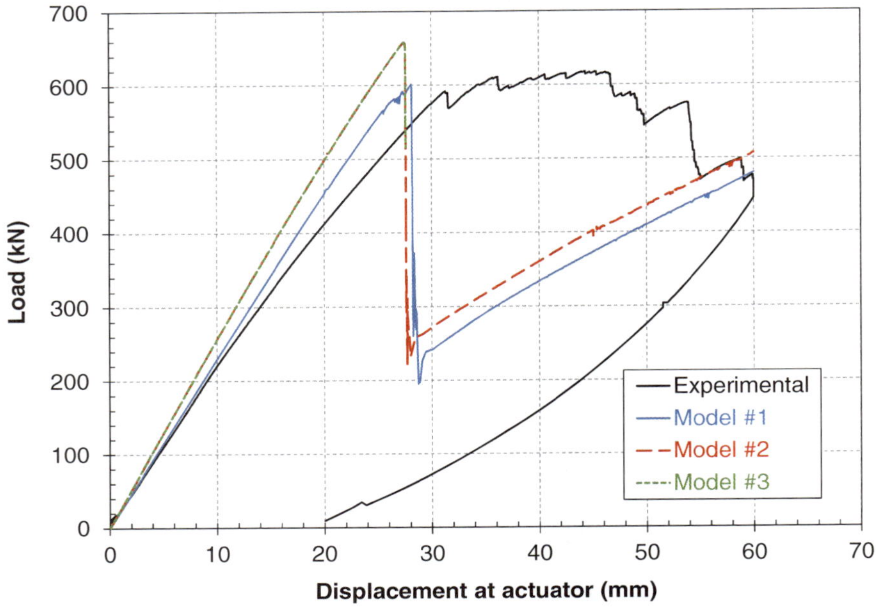

Fig. 3 Global load-deflection behavior from experimental testing and FEA

It is evident that the stiffness of the hybrid bridge deck system was best represented by Model #1, where no internal UHPC stiffener blocks were included. The other two FEMs that included UHPC stiffener blocks showed nearly identical load-deflection curves, while exhibiting stiffer behavior than experimental results. It appears that the addition of tube-to-tube elements, restricting the movement of the horizontal GFRP rods and PT GFRP rods within a specific transverse path through the hybrid bridge deck system, does not exert any discernible impact on the global behavior of the FEM.

Focusing on the conditions at peak condition, Model #1 reached a peak load of 602 kN at a deflection of 28.2 mm whereas both Models #2 and #3 attained a peak load of 656 kN with a deflection of 27.6 mm. Thus, the results from all three developed FEMs provided an overestimation of the initial peak load for the hybrid bridge deck system, when compared with experimental results. With attention to the post-peak behavior, significant discrepancies between the experimental results and the FEA results from all three models were observed. The post-peak load resistances calculated from Model #1, #2 and #3 were 198, 232 and 232 kN, respectively. This is compared with a much smaller decrease in load resistance observed experimentally, where the post-peak load resistance was 570 kN. After stabilization of load resistance, however, the recovery zone in load-deflection curves obtained from the FEMs show a similar stiffness as the experimental data.

6.2 Comparison of Local Performance

Due to the lack of apparent difference in global behavior between Models #2 and #3, where tube-in-tube elements were used in the latter model, closer investigation of differences in local behavior was conducted. The first step is a comparison of the tensile strain at the bottom face of the CFRP sheet at mid-span. The load-strain behavior obtained from experimental testing and FEA is provided in Fig. 4.

Once again, the FEA curve that most closely approximates the experimental data was obtained from Model #1; however, in the case of local tensile behavior in the CFRP sheet, an observable difference in performance is detected between Models #2 and #3, where slightly less tensile strain was obtained from Model #2. This set of local behavior comparison does confirm that the soundness of the developed FEM in representing the overall performance of the hybrid bridge deck system.

Next, the differential vertical displacement between the location of the ducts and the horizontal PT GFRP rod is compared between the three FEMs. In experimental testing, the GFRP rod moved solely through interior of the ducts, which traversed the interior of the GFRP hollow box beams; however, this condition was not imposed in Models #1 and #2. The absolute value of the differential vertical displacement with increased loading up until peak load is presented in Fig. 5 for the three developed FE models.

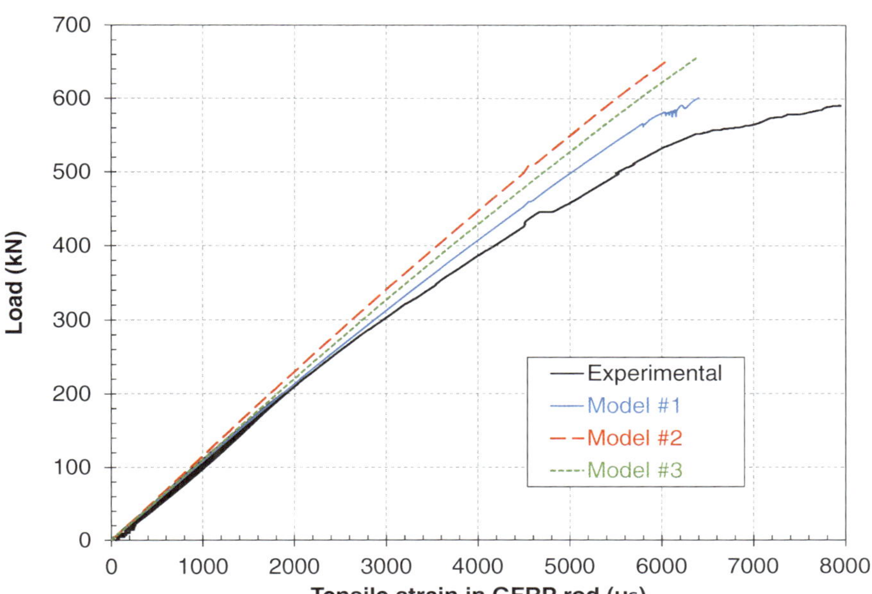

Fig. 4 Load-strain behavior in the CFRP sheet from experimental and FEA testing

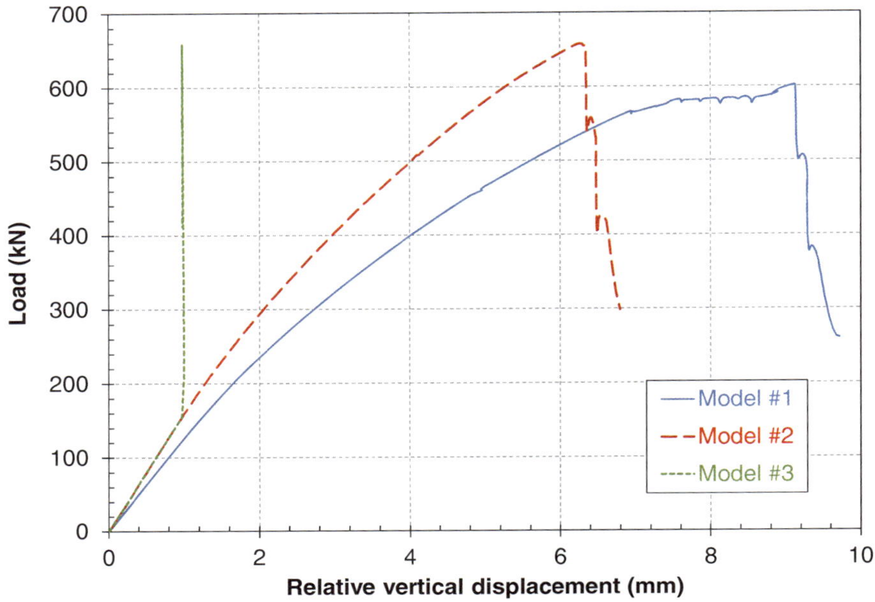

Fig. 5 Comparison of differential movement between the GFRP rods and rod ducts

As expected, the relative vertical displacement in Model #3 was limited to a maximum value of 1 mm, due to the initial imposed clearance of 1 mm between the outer diameter of the GFRP rod and inner surface of the duct. More interestingly, it is observed that Models #1 and #2 reached a maximum differential displacement of approximately 9.0 and 6.4 mm, respectively, at peak condition. It is clear that the increased global stiffness of Model #2 also resulted in a decreased level of differential movement. Lastly, in regards to Model #2 and #3, it is clear that, even though significantly greater differential displacement was obtained between the GFRP rod and the actual location of the ducts in Model #2, there was no influence on the global behavior observed. To investigate the cause behind this phenomenon, both the load-strain behavior and the strain distribution data along the GFRP rod at peak condition are presented in Figs. 6 and 7, respectively.

The tensile strain at peak condition for Models #1, #2 and #3 were 410, 813 and 1004 με, respectively. Compared with a strain value exceeding 3500 με obtained through experimental testing, all of the FEM configurations explored in this paper greatly underestimated the tensile strain in the GFRP rods, within a margin of 3.5–9 times less than the experimental value. Even in the case of Model #3, where tube-in-tube elements were used to allow free movement through the bridge system, it appears that an additional unknown constraint is limiting the elongation of the GFRP rod. It is clear by looking at the strain distribution along

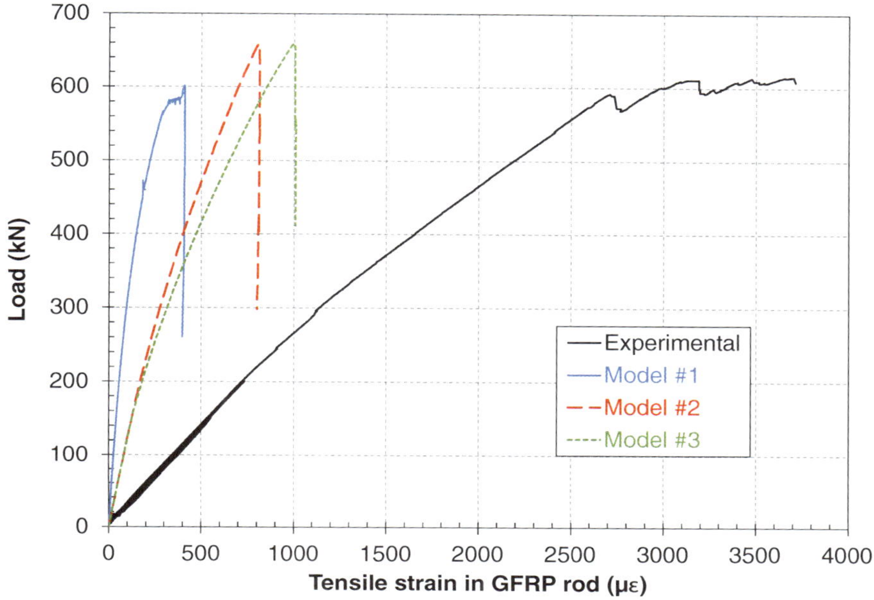

Fig. 6 Comparison of tensile strain in GFRP rods

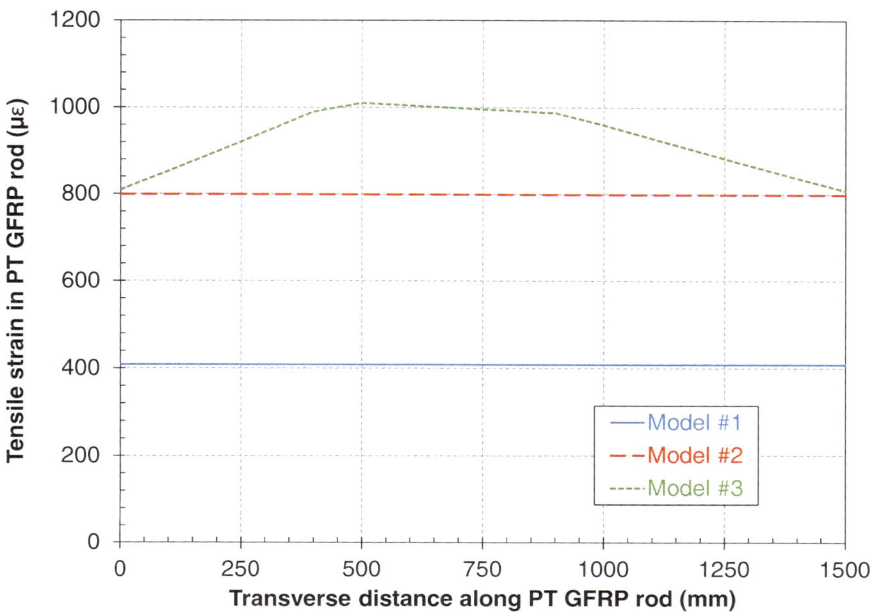

Fig. 7 Comparison of strain distribution along the GFRP rod

Table 5 Summary of computation time taken FEA

Displacement (mm)	Computation time (s)		
	Model #1	Model #2	Model #3
2	512	1428	981
4	302	510	599
6	438	507	526
8	523	508	521
10	608	454	69
12	654	798	844
14	1039	1184	1168
16	1079	2061	1629
18	1371	1872	1813
20	1546	2275	2131
22	3273	11,899	9837
24	1806	32,591	25,393
26	6345	3345	4520
28	15,859	37,489	22,690*
30	29,725	7455	–
Total	65,080	104,376	72,721

*did not complete load step

the GFRP rod that accurate representation of transverse curvature in the GFRP rod was depicted solely in the case of Model #3, whereas constant tensile strain levels were obtained from Models #1 and #2.

6.3 *Evaluation of Required Computation Power*

A detailed breakdown of the computation time required for each load step, divided into displacement increments of 2 mm, is provided in Table 5 for the three FEMs.

Clearly, the model that required the most time to complete was Model #2, even taking into consideration the fact that Model #3 did not complete the final step (to a displacement of 30 mm) due to numerical non-convergence. It does appear, however, that the more accurate behavior (in terms of local behavior in the CFRP sheet and GFRP rods) can be obtained with less computational cost.

7 Conclusions

Examination of the most appropriate (in terms of accuracy and required resources) FEA techniques to apply when modelling the behavior of the specially designed hybrid bridge deck system presented several interesting results. These include:

1. Reasonable accuracy of global peak behavior was obtained from all three FEMs examined, with greater stiffness provided in models incorporating the presence of internal UHPC stiffener blocks. Overestimation of the peak load, compared with experimental results, within a range of 2–12 % was obtained. Significantly greater loss in load resistance was obtained in the FEMs though similar trends in the progressive failure zone were observed.
2. Validation of local behavior (through comparison of tensile strains in the CFRP sheet), showing good correlation with experimental results for all FEMs.
3. Significant underestimation (within 3.5–9 times less than experimental results) of tensile strain in the horizontal GFRP rod was obtained from all FEMs studied. The FEM utilizing tube-to-tube elements (Model #3) to define the transverse movement path of GFRP rods within the hybrid bridge system was the only model to provide transverse curvature behavior within the GFRP rod.
4. Analysis of computational time required to complete FEA show that Model #3 allowed for the best optimization of computational resources and accuracy of global and local performance.

From the results of this study, Model #1 is recommended for use if characterization of the general global behavior, including estimating of peak conditions, is required. For future study and refinement of the FEM, it is recommended that Model #3 (where UHPC stiffener blocks and tube-to-tube elements were implemented) be used. Additional calibration of parameters are required to identify the unknown constraint placed on the GFRP rods, to allowing greater tensile strains to be expressed in these structural elements.

Acknowledgments The authors would like to thank Lafarge Canada, Sika Canada and Tyfo Inc. for their generous donation of materials used. We would also like to acknowledge the University of Calgary and the Natural Sciences and Engineering Research Council of Canada (NSERC) for their financial support. Lastly, we would like to thank the technical staff for their invaluable help during experimental testing.

References

1. Bank LC, Oliva MG, Russell JS, Dieter DA, Dietsche JS, Berg AC, Ehmke FG, Carter J, Hill R, Henke R, Gallagher B (2005) Bridge B-20-133 on US_151 with fiber reinforced polymer reinforced concrete deck. *FHWA report no. IBRC-001*. Wisconsin DOT, Madison
2. Hassan T, Rizkalla R (2003) Investigation of bond in concrete structures strengthened with near surface mounted carbon fiber reinforced polymer strips. J Compos Constr 7(3):248–257
3. Cheng L, Karbhari VM (2006) New bridge systems using FRP composites and concrete: a state-of-the-art review. Prog Struct Mat Eng 8(4):143–154
4. El-Hacha R, Chen D (2012) Behaviour of hybrid FRP-UHPC beams subjected to static flexural loading. Compos B Eng 43(2):582–592
5. Chen D, El-Hacha R (2014) Flexural performance of hybrid multi-girder bridge deck with all-GFRP transverse composite action system. In: Proceedings of 16th European conference on composite materials. Seville, Spain, June 22–26, p 8
6. Canadian Standards Association (2006) Canadian highway bridge design code (CSA S6-06). Mississauga, CSA

7. Strongwell (2013) FIBREBOLT® properties. Strongwell, Bristol
8. Chen L, Graybeal BA (2010) Finite element analysis of ultra-high performance concrete: modeling structural performance of an AASHTO type II girder and a 2nd generation Pi-girder. FHWA publication no. FHWA-HRT-11-020. FHWA: McLean
9. Chen D, El-Hacha R (2014) Cohesive fracture study of a bonded coarse silica sand aggregate bond interface subjected to mixed-mode bending conditions. Polymers 6(1):12–38
10. Chen D, El-Hacha R (in press) Investigation of hybrid FRP-UHPC beams under flexure with finite element methods. ACI Special Publication

Free Vibration Analysis and Seismic Performance Assessment of Two Approach Viaducts of Bosphorus Suspension Bridge

Selcuk Bas, Nurdan M. Apaydin and Zekai Celep

Abstract The current study aims to determine the dynamic vibration characteristics and structural earthquake performance of the approach viaducts of Bosphorus Suspension Bridge. Ortaköy viaduct at European continent and Beylerbeyi viaduct at Asian continent have a length of 231 and 235 m, and five spans and four spans, respectively. The viaducts are side span of the bridge and are supported at the base instead of hanger elements. Based on the project specifications of the viaducts, 3-D detailed finite element model of the viaducts is developed by the use of frame elements, and natural frequencies and corresponding mode shapes of the viaducts are presented in the study. Paying attention to these considerations, pushover (POA) and nonlinear time-history analysis (NTHA) are performed to attain structural earthquake performance of the viaducts according to Turkish code for the earthquake design of railways bridges (TSC-R/2008) and Caltrans (CALTRANS-2001) seismic design of steel bridge. Accordingly, the earthquake performance of the bridge is presented for each non-linear analysis method considering these codes.

S. Bas (✉) · Z. Celep
Structural and Earthquake Engineering Department,
Istanbul Technical University, Istanbul, Turkey
e-mail: selcukbas@itu.edu.tr; sbas@bartin.edu.tr

Z. Celep
e-mail: celep@itu.edu.tr

S. Bas
Civil Engineering Department, Bartin University, Bartin, Turkey

N.M. Apaydin
Division Directorate, General Directorate of Turkish State Highways, Istanbul, Turkey
e-mail: napaydin@kgm.gov.tr

A. Caner et al. (eds.), *Developments in International Bridge Engineering*,
Springer Tracts on Transportation and Traffic 9,
DOI 10.1007/978-3-319-19785-2_6

1 Introduction

The observed considerable bridge failures after from the catastrophic earthquakes such as 1989 Loma Prieta and 1994 Northridge earthquakes in the United States and 1999 Kocaeli earthquake in Turkey have resulted in growing interests for seismic safety evaluation of the existing bridges. For this purpose of evaluation, certain nonlinear response analysis methods considering performance-based assessment have been developed, such as nonlinear static and dynamic analysis methods. In literature, studies and guidelines are more concentrated on the seismic assessment of buildings rather than bridges. Bridge researchers and engineers are currently investigating similar concepts and procedures in order to develop simplified procedures for performance-based seismic evaluation of bridges. Concerning this point, Aydinoglu and Onem [1] were applied IRSA [2] method to a set of bridges in order to evaluate the reliability of the method by comparing its results with those obtained from nonlinear response history analysis. They remarked that IRSA gave the relatively accurate results on the prediction of bridge nonlinear response demand. Barron [3] and Shinozuka et al. [4] implemented the nonlinear static method for determination of the fragility curves for bridges. In these studies, the capacity spectrum method generally was used to estimate the displacement demand of bridges.

With regard to the existing critical bridges in Turkey, particularly in Istanbul such as Bosphorus and Fatih Sultan Mehmet Suspension Bridges, Apaydin [5] studied on the determination of structural earthquake performance of Bosphorus and Fatih Sultan Mehmet Suspension Bridges. In this study, free vibration analysis of the bridges was also performed by developing the sophisticated 3D finite element model of the bridges. Besides, the Bosphorus Bridge is modeled by considering the approach viaducts which are the side span of the bridge. It is concluded that the first dominant mode period of the bridge with the approach viaducts is $T_1 = 13.49$ s in the lateral direction. Kosar [6] also developed the detailed 3D finite element model of the bridge without the approach viaducts to compare the results from free vibration analysis and those attained by experimental study of AVS. The study demonstrated that the fundamental mode period of the bridge without the approach viaducts is $T_1 = 14.51$ s in the lateral direction. From the analytical studies performed by Apaydin [5] and Kosar [6], it is concluded that there has been a considerable difference in the fundamental mode period between three-dimensional model of Bosphorus Suspension Bridge without and with approach viaducts.

Regarding this difference between two models as significant, it is aimed in the recent study to determine dynamic response characteristics and structural earthquake performance of the approach viaducts of Bosphorus Suspension Bridge in Istanbul [7]. In the scope of the study, free vibration analysis of the approach viaducts is firstly carried out to attain the fundamental modes and the associated mode shapes of them. Taking these dynamic modal specifications into account, seismic performance assessment of the viaducts is determined by using pushover method and nonlinear time-history method. Concerning the procedure of these nonlinear analysis methods, the requirements of Turkish Seismic Code for Railways (TSC-R/2008) [8] are considered. In an effort to be complementary for TSC-R/2008, the provisions of Turkish Seismic Code (TSC-2007) [9] to be built in earthquake zones are also utilized. Finally, structural earthquake performance of the viaducts according to TSC-R/2008 and Caltrans (CALTRANS-2001) [10] are specified by using two non-linear analysis methods, and the results obtained from these non-linear analysis are presented.

2 Description of the Approach Viaducts

The Bosphorus Bridge has two approach viaducts which are the side span of the bridge. As shown in Fig. 1a, b, one of them at the western portion of the city in Europe called as Ortakoy has a length of 231 m and five spans, and the other one at the eastern portion in Asia called as Beylerbeyi has a length of 255 m and four spans. Additionally, the viaducts have no hanger elements, and thus have different columns supported at the base and viaduct superstructure.

Superstructure of the viaducts is made up of steel lateral girder beams located on the deck with 3.0 m span along the viaducts and steel hollow beam carrying concrete and asphalt pavement layers and traffic loads. Steel hollow beams restrained by columns having steel circular box section are continuous along the viaducts. The complete section of the superstructure of the viaducts is shown in Fig. 2. Besides, there is asphalt pavement with the height of 38.0 mm on the highway lanes. In the analysis, weight of concrete deck and asphalt pavement on the lanes are regarded as dead load, and the effect of weight of these parts on the structural stiffness and performance of the viaducts is also taken into consideration. Vertical stiffness of deck cross section of the viaducts is much lower than lateral stiffness, but they are relatively low enough when comparing to stiffness of the steel hollow beam of the viaducts.

Fig. 1 General view of Ortakoy viaduct (**a**) and Beylerbeyi viaduct (**b**)

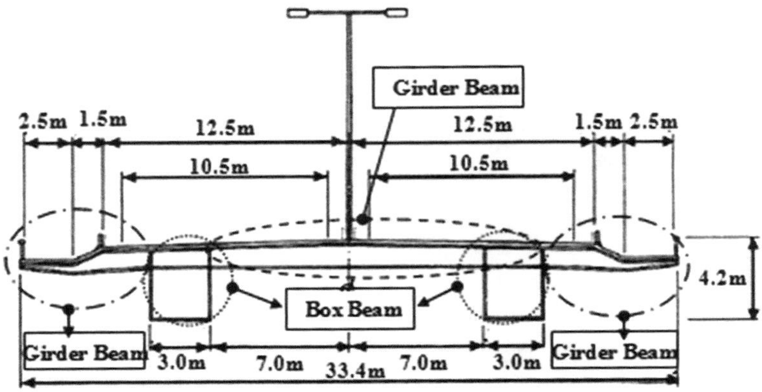

Fig. 2 Superstructure section of the viaducts [11]

The steel hollow beams of viaducts are supported by the columns. The cross sectional dimension of these beams whose sections are braced with steel plate elements in longitudinal direction is 3.00 m × 3.80 m. Owing to the fact that top and bottom flange, web thickness and height of bracing plate elements of the steel hollow beams are variable along the viaducts, the maximum cross sectional dimensions of the steel hollow beams are considered in 3-D structural finite element model. Accordingly, the need to increase the dimensions of the hollow beam section at the points of stress concentration is relatively taken into consideration. The columns of the viaducts with pipe section have different wall thickness and different height in each bent of the structural system, and also support conditions at the bottom and top ends of them as shown in Fig. 3. The section properties of the viaducts are considered according to the project specifications upon developing their finite element models.

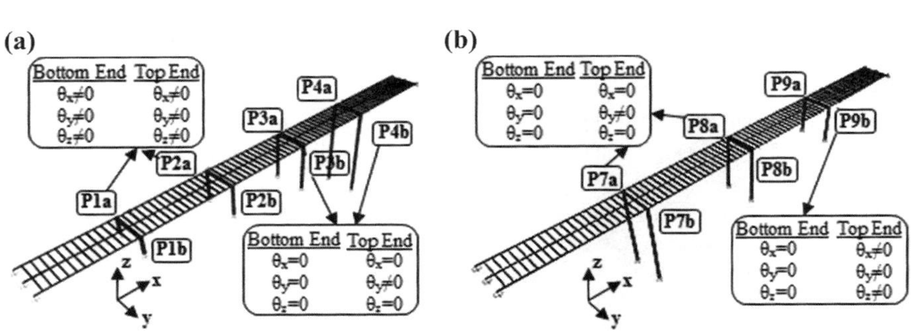

Fig. 3 Support conditions of Ortakoy (**a**) and Beylerbeyi (**b**) approach viaducts

3 Dynamic Vibration Analysis of the Viaducts

3.1 Finite Element Model of the Viaducts

The detailed 3-D finite element computational model of the viaducts is developed by using SAP2000 [12]. All structural members of the viaducts, steel deck, steel hollow beams, lateral girder beams and the pipe columns are taken into consideration as frame element in the model. Hollow beam has a pinned support having releases in longitudinal direction at the tower point due to the expansion joint, and has pinned support due to the landside. Based on the project drawings, these specifications are considered as the same for two viaducts when developing the computational model of the viaducts. In Fig. 4, all support conditions at these points are shown on the

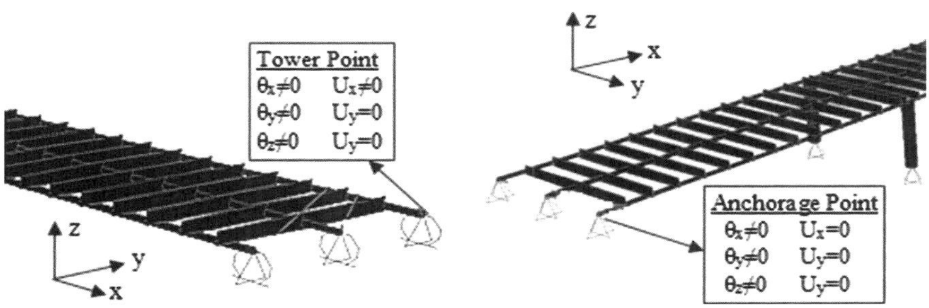

Fig. 4 The support conditions of anchorage and tower points of the viaducts

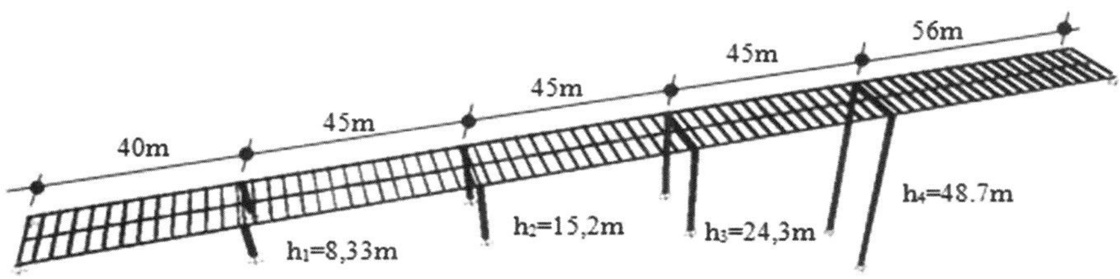

Fig. 5 Finite element model of Ortakoy approach viaduct

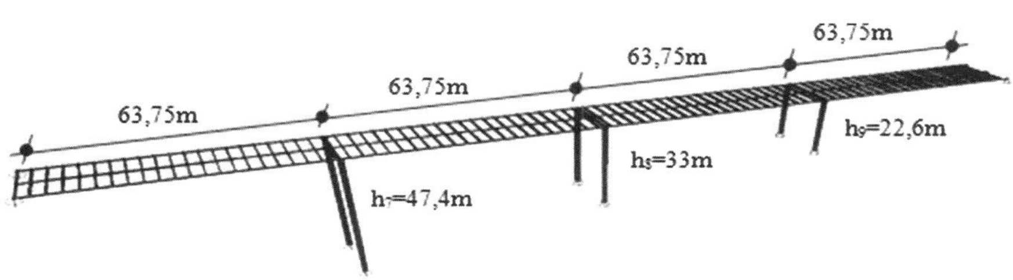

Fig. 6 Finite element model of Beylerbeyi approach viaduct

finite element model of the viaducts. Accordingly, 3D finite element model of the viaducts is constituted, and the completed finite element model of them is demonstrated in Figs. 5 and 6 for Ortakoy and Beylerbeyi viaducts, respectively.

3.2 Structural Dynamic Characteristics of the Approach Viaducts

Free vibration analysis of the viaducts is performed by means of SAP2000. From the analysis, the effective mode shape of the viaducts is determined in transverse direction. The results of dynamic parameters and amplitudes of the fundamental mode shape of the viaduct are given in Table 1. For pushover analysis (POA) of

Table 1 Results of the free vibration analysis of the viaducts

Dynamic vibration characteristics of the viaducts						
Viaduct	Amplitudes of dominant mode shape	T_{1y} (s)	f_{1y} (cyc./s)	ω_{1y} (rad/s)	M_{y1} (kN · s²/m)	$\Gamma y1$
Ortakoy	$\Phi_{11} = 0.26$	1.27	0.79	4.95	5368	1.27
	$\Phi_{12} = 0.75$					
	$\Phi_{13} = 1.00$					
	$\Phi_{14} = 0.79$					
Beylerbeyi	$\Phi_{17} = 0.84$	1.55	0.65	4.05	6305	1.23
	$\Phi_{18} = 1.00$					
	$\Phi_{19} = 0.46$					

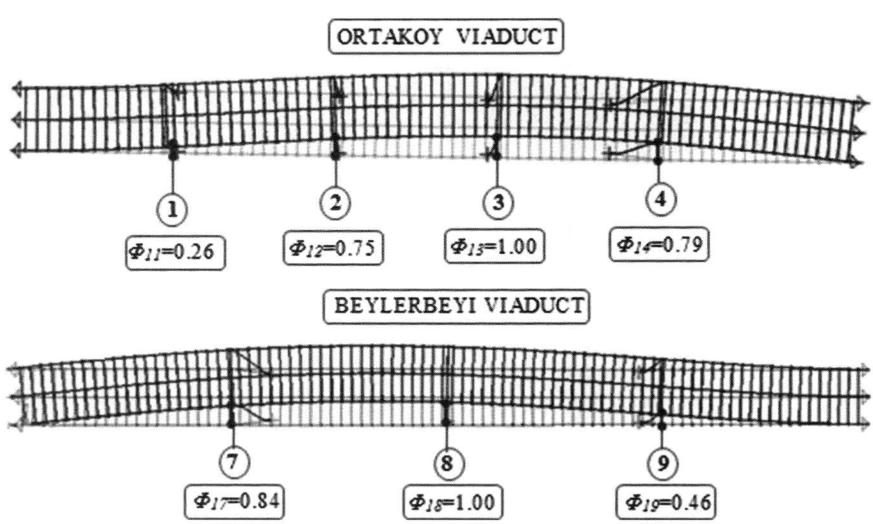

Fig. 7 Schematic presentation of the fundamental mode shape of the viaducts

the viaducts, this mode shape is taken into consideration. As can be also seen from Fig. 7, the monitoring point of the viaducts which is necessary especially for POA is selected as Bent no: 3 and Bent no: 8 for Ortakoy and Beylerbeyi viaducts respectively, so that these points can be at the mass center of the superstructures of the viaducts.

3.3 Simulated Earthquake Ground Motion

In order to perform NTHA, three simulated earthquake actions are taken into account for maximum demand. Accordingly, Loma Prieta (NW, 1989), Erzincan (NE, 1992) and Kocaeli (NE, 1999) earthquake records attained from the Pacific Earthquake Engineering (PEER) strong motion database are considered so as to produce new earthquake records by using the program of Seismo-Match [13]. For

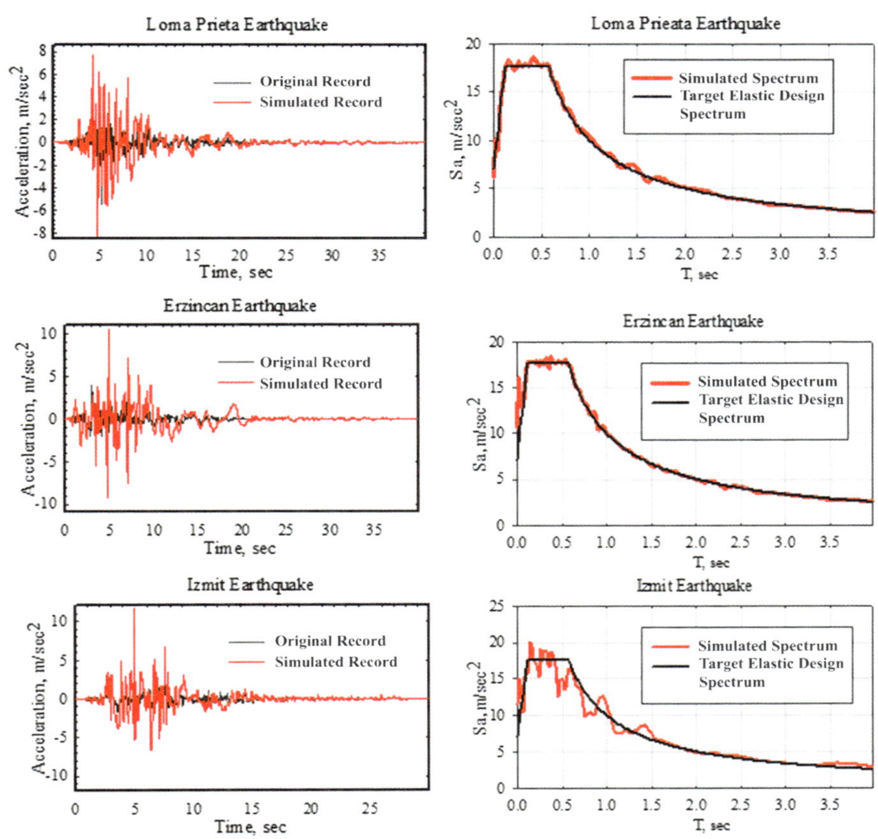

Fig. 8 Simulated earthquake strong ground motion records

this purpose, these earthquake actions are simulated by considering the compatibility of them to target elastic design spectrum whose all specifications are given in TSC-R/2008. In Fig. 8, the original and simulated earthquake records are shown by their target elastic design spectrum.

4 Earthquake Performance of the Approach Viaducts

For performing POA, material non-linearity of the structural members of the viaducts such as hollow beams and pipe columns are regarded as cumulative plastic deformation, commonly well-known plastic hinge. The possible plastic hinges are limited to the columns and hollow beams. For these potential members of the viaducts, the behavior of plastic hinges is represented as moment-curvature relation by considering the strain-hardening effect. Following the completion of these preparations, fundamental mode-based pushover analysis is performed by using SAP2000. From this analysis, the pushover curve is determined for each viaduct as shown in Fig. 9.

The obtained pushover curves of the viaducts are then transformed to capacity curve. Thus, the seismic performance point of the viaducts which presents the spectral displacement demand is obtained by applying the equal displacement rule due to the fact that the capacity curve intersects elastic acceleration-displacement response spectrum (ADRS) at the long-period range. Spectral displacement and acceleration at the performance point of the viaducts are shown in Fig. 10. After determination of the plastic deformation of each element, the sectional and structural earthquake performance level of the approach viaducts are determined by depending on calculated total curvatures. The earthquake performance of Ortakoy viaduct at the level of section and structure is given in Table 2 only for the column at Bent no: 3 because no damage takes place at the other columns and hallow beam. Due to fact that Beylerbeyi viaduct has no damage, any knowledge for its performance is not given in the table.

Consequently, Beylerbeyi viaduct is directly provides Operational (OP) performance level. When it comes to Ortakoy viaduct, Fully Operational (FOP)

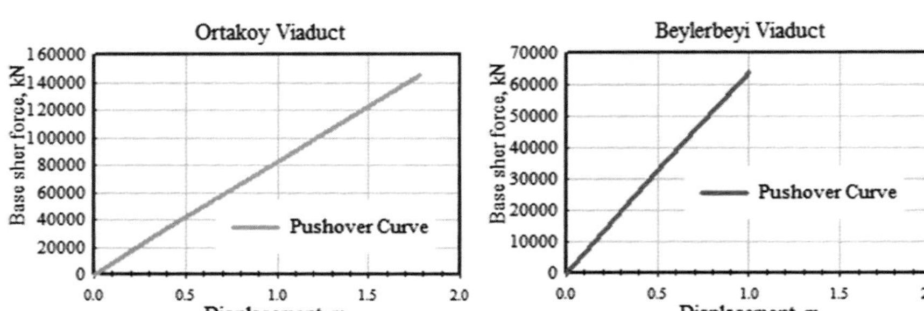

Fig. 9 Pushover curves of the viaducts

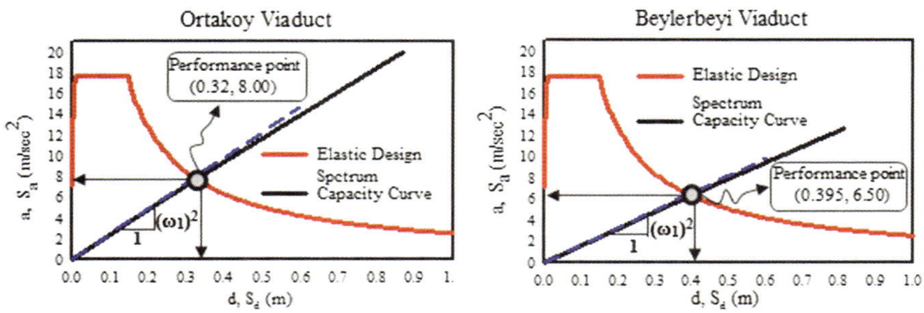

Fig. 10 Earthquake performance point of Ortakoy and Beylerbeyi approach viaducts

Table 2 Sectional and structural earthquake performance of Ortakoy viaduct (POA)

Ortakoy viaduct								
Bent No.	Pier No.	Pier	Max. total curvature (1/m)	Max. total strain	Strain limit		Sectional perfor-mance	Structural performance
					Minimum (MN)	Repairable (RP)		
TSC-R, 2008								
3	P3a	P3a-i	0.0059	0.0061	<0.008	<0.025	MN	Fully operational
		P3a-j	0.0063	0.0063			MN	
	P3b	P3b-i	0.0062	0.0062			MN	
CALTRANS, 2001								
3	P3a	P3a-i	0.0059	0.0061	<0.003	<0.008	RP	Operational
		P3a-j	0.0063	0.0063			RP	
	P3b	P3b-i	0.0062	0.0062			RP	

Table 3 Sectional and structural earthquake performance of Ortakoy viaduct (NTHA)

Ortakoy viaduct								
Bent No.	Pier No.	Pier	Max. total curvature (1/m)	Max. total strain	Strain limit		Sectional perfor-mance	Structural perfor-mance
					Minimum (MN)	Repairable (RP)		
TSC-R, 2008								
3	P3a	P3a-i	0.0072	0.0065	<0.008	<0.025	MN	Fully operational
		P3a-j	0.0051	0.0051			MN	
	P3b	P3b-i	0.0061	0.0057			MN	
		P3b-j	0.0057	0.0057			MN	
CALTRANS, 2001								
3	P3a	P3a-i	0.0072	0.0065	<0.003	<0.008	RP	Operational
		P3a-j	0.0051	0.0051			RP	
	P3b	P3b-i	0.0061	0.0057			RP	
		P3b-j	0.0057	0.0057			RP	

performance level is provided according to the requirements of TSC-R/2008 while Operational (OP) performance level is provided by relying on the basics of CALTRANS-2001. As a result of implementation of NTHA, it is determined from the analysis that the structural members of Beylerbeyi viaduct have no plastic deformation. So, Beylerbeyi viaduct directly provides Fully Operational (FOP) performance. In Table 3, the sectional and structural earthquake performance of Ortakoy viaduct are summarized. According to the requirements of TSC-R/2008 and CALTRANS-2001, the structural earthquake performance of Ortakoy viaduct is determined as Fully Operational (FOP) and Operational (OP), respectively.

5 Conclusion

Dynamic vibration characteristics and structural earthquake performance of the approach viaducts of Bosphorus Suspension Bridge are investigated in the current study. Ortakoy and Beylerbeyi approach viaducts have different structural system when compared to the suspension bridge. Based upon the project specifications for the viaducts, this can be explained that the viaducts are carried by the columns supported at the base instead of the suspenders, and that they have different support conditions at the anchorage and tower points which are expansion joint of the suspension bridge.

As a result of modal analysis of the viaducts, the fundamental mode period and corresponding mode shapes of the viaducts are obtained in transverse direction. The reason for determination of this type of fundamental mode shape is based upon that the viaducts have pinned support conditions at the anchorage and tower points shown in Fig. 4. Additionally, the effectiveness of the bending rigidity on the hollow beams and the columns in transverse direction allows the viaducts to have lateral mode shape. It is also observed that there is a longitudinal mode shape, but it is not as effective as the lateral mode shape. This conclusion can be explained by the effectiveness of the axial rigidity on the hollow beams in longitudinal direction. In Table 1, the results from the modal analysis of the viaducts are summarized only for their effective mode shapes schematically demonstrated in Fig. 7. Accordingly, maximum translation amplitude of Ortakoy and Beylerbeyi viaducts is determined as the Bent no: 3 and Bent no: 8, respectively. Upon implementing pushover analysis for performance assessment of the viaducts, these nodes are selected as monitoring points.

To be able to determine the seismic performance of the viaducts, push over analysis (POA) and nonlinear time-history analysis (NTHA) are performed. It is observed from the pushover curve of the viaducts that the response of the viaducts to incrementally increasing earthquake load is approximately quasi-linear as shown in Fig. 9. The main reason for this type of capacity curve is due to the fact that the viaducts have higher degree of freedom. Moreover, the support conditions at expansion joints as shown in Fig. 4 and of the columns as shown in Fig. 3 are also effective on such elastic behavior of the viaducts. Based on sectional performance of the structural members of the viaducts, Ortakoy viaduct meets the requirements of TSC-R/2008 with Fully Operational (FOP) performance level whereas Operational (OP) performance level is provided according to the provisions of CALTRANS-2001. Beylerbeyi viaduct provide directly Fully Operational (FOP) performance level for both codes due to fact that all structural members of the viaduct have no damage.

As a consequence of NTHA, taking into sectional performance of the viaducts into account, Ortakoy viaduct provides Fully Operational (FOP) performance level for TSC-R/2008 and Operational (OP) performance level for CALTRANS-2001 as shown in Table 3. As to Beylerbeyi viaduct, the structural earthquake performance of the viaduct is determined as Fully Operational (FOP) level for both codes because of having no damage. Accordingly, it can be said that the viaducts provides the same structural earthquake performance level for POA and NTHA. Nevertheless, the damage obtained from NTHA distributes uniformly on the column P3 when compared to those attained from POA.

References

1. Aydinoglu MN, Önem G (2007) Nonlinear performance assessment of bridges with incremental response spectrum analysis (IRSA). In: ECCOMAS thematic conference on computational methods in structural dynamics and earthquake engineering, Rethymno, Greece
2. Aydınoğlu MN (2003) An incremental response spectrum analysis based on inelastic spectral displacements for multi-mode seismic performance evaluation. Bull Earthq Eng 1(1):3–36
3. Barron R (1999) Spectral evaluation of seismic fragility of structures. Dissertation, Department of Civil Structural and Environmental Engineering, State University of New York at Buffalo
4. Shinozuka M, Feng MQ, Kim H, Kim S (2000) Nonlinear static procedure for fragility curve development. J Struct Eng ASCE 126(12):1287–1295
5. Apaydın N (2010) Earthquake performance assessment and retrofit investigations of two suspension bridges in Istanbul. J Soil Dyn Earthq Eng 30:702–710
6. Kosar U (2003) System identification of Bosphorus Suspension Bridge. MSc thesis, Bosphorus University, Istanbul, Turkey
7. Bas S (2011) Dynamic analysis and structural earthquake performance of two approach viaducts of the Bosphorus Bridge. MSc thesis, Istanbul Technical University, Istanbul, Turkey
8. Seismic technical specifications for coastal and harbour structures, railways and airport construction, chapter-3: design essentials of railway bridges under earthquake action, Seismic code for railway-(TSC-R) (2008) Ministry of Transportation, Ankara, Turkey
9. Turkish Seismic Code-TSC, specification for buildings to be built in earthquake zones (2007) Ministry of Public Work and Settlement, Ankara, Turkey
10. Guide specifications for seismic design of steel bridges (2001) State California Department of Transportation (CATRANS), Sacramento, CA, USA
11. The Bosporus Bridge approach viaducts drawings (1968) Consulting engineers. Freeman, Fox and Partners, Westminster, London, UK
12. Integrated structural finite element analysis and design of structures (SAP2000) (2011) Computers and Structures Inc., Berkeley, CA, USA
13. Software applications for analysis of structures subjected to seismic actions (2010) Seismosoft Ltd, Via Boezio 10, Pavia, Italy

Turkish Solar Radiation Zones for Segmental Box Girder Bridge Design Under Temperature Gradient Loading

A.I. Yilmaz, U. Albostan, A. Caner and O. Kurc

Abstract Solar radiation developing non-linear temperature distribution at the depth of the concrete structures may cause significant stress changes along the long span box type bridges. Segmental concrete box bridges are one of the example in which stresses change not only due to annual uniform temperature changes but also due to daily temperature changes that can develop a second type of thermal difference; namely, thermal gradient, which forces to have internal thermal stresses within a cross-section. Over the years it has been observed that nonlinear temperature distribution develop at the depth of box girder type bridges can cause bending moments as high as the ones generated due to live loads that may result in concrete cracking. American Highway Association of Transportation Officials (AASHTO) bridge specifications recognize the use of thermal gradient loads in design since 1989. In this specification, the U.S. is divided into 4 zones per the country's solar radiation zones and gives some gradient values to be applied through the depth of the girder. The aim of this study is to construct a simple similar solar radiation map for Turkey to be used in design of segmental bridges. Temperature and solar radiation data from eight cities in different regions of Turkey are collected, and used in analysis of a box girder bridge model through a thermal finite element analysis program. Thermal differences through the depth of the girder are determined to obtain the design gradient values for the bridge that

A.I. Yilmaz (✉) · U. Albostan · A. Caner · O. Kurc
Department of Civil Engineering, Middle East Technical University, 06800 Ankara, Turkey
e-mail: arzuipekyilmaz@gmail.com

U. Albostan
e-mail: albostan@metu.edu.tr

A. Caner
e-mail: acaner@metu.edu.tr

O. Kurc
e-mail: kurc@metu.edu.tr

A. Caner et al. (eds.), *Developments in International Bridge Engineering*,
Springer Tracts on Transportation and Traffic 9,
DOI 10.1007/978-3-319-19785-2_7

can be hypothetically constructed in different parts of Turkey. The end results are used to develop a recommendation for a simple solar map that is related to thermal gradients to be used in design of segmental bridges.

Keywords Thermal gradient · Solar radiation · Segmental bridges · Box girder bridges · Design codes

1 Introduction

Segmental concrete box bridges are subject to not only the annual uniform temperature changes—which can be accommodated by providing expansion joints, flexible bearings or low lateral stiffness substructures—but also a second type of thermal variation which occurs throughout a cross section, named, thermal gradient. Solar radiation induced temperature differences along the depth of bridge cross section are mainly due to convection from surroundings. Conduction cause nonuniform expansion and contraction of a bridge. This change may affect bridge section in two ways. Under positive thermal gradient; which is the condition of deck warmer than webs, the top surface of the structure will expand more than the bottom surface, causing the structure to deflect upward, and under negative thermal gradient, the case of the top of girder is colder than web, the tendency to contraction will cause tensile stresses to develop at deck which cause cracking type of problems. Moreover; since the thermal gradient profile is nonlinear, compatibility stresses are generated to satisfy the assumption of Bernoulli beam theory that plane sections remain plane. Furthermore; boundary conditional restrains cause axial and bending stresses to develop in the cross section.

Positive gradient gets higher when several days of cool weather is followed by unclouded warm days with intense solar radiation and light winds. A maximum negative gradient takes place when several days of warm weather is followed by a severe cold weather and rain cooling the deck occurs. The cause of large gradients is a result of low conductivity of concrete that the heat gain cannot be transferred quickly to other parts of the cross section.

The principal causes of vertical temperature differences in a girder can be summarized into 3 titles: Climatological factors, geometrical and material properties. Primary climatological factors; solar radiation, air temperature, and wind speed depends on geographical parameters such as latitude, longitude, altitude and also time. Geometric factors are the differences in cross-sectional geometry, overlay thickness, and orientation of bridge. Finally the material properties are the thermal conductivity, color, density, specific heat, and absorptivity of bridge deck components.

Ignoring thermal gradients in design may cause the concrete cracking type of problem. The Fourth Danube Bridge in Vienna, the Newmarket Viaduct in New Zealand, and The Jagst Bridge in Untergreisheim, Germany [13] can be given as examples for this situation. In Colorado; four cast-in-place segmental prestressed concrete box girder bridges were subjected to cracking in the webs and bottom deck soffits which was imputed to thermal gradients [14]. Csagoly and Bollman observed significant opening of segmental bridge dry joints due to thermal gradients in a variety of bridges studied in the Florida Keys [23]. Leonhardt et al. [17] reported 5 mm cracks in the Jagst Bridge in Germany which mainly attributed to differential temperatures. According to Wood [25], in a continuous structure thermal gradient stresses caused by longitudinal moments can equal those of live loads.

A solar radiation map that is related to thermal gradient for concrete structures is not available in Turkey. The aim of this pioneering study is to relate the solar radiation effects to the thermal gradient for segmental concrete box bridges to be used in Turkish bridge designs.

2 Previous Studies, Design Codes and Specifications Worldwide

In 1967, German Code DIN 1072 [12] was considering a differential linear temperature distribution decreasing 5 °C from upper surface to the soffit. New Zealand Code's recommendation (1970) considering a constant temperature in the top slab replaced by a highly nonlinear-sixth order curve in 1973; and once again changed into the requirement to use a fifth order parabola for first 1200 mm depth of girder with a 57.6 °F (32 °C) temperature gradient in 1978 as Priestly cites [22].

The first recommendation to consider thermal gradient in segmental bridge design in the United States was done by Prestressed Concrete Institute-Post Tensioning Institute (PCI-PTI) in 1977 as a constant gradient in top slab with 18 °F (10 °C) temperature differential [20]. After that Hoffman, McClaure and West made an experimental study in Pennsylvania and proposed to use 18 °F (10 °C) differential, instead of PCI-PTI's 18 °F (10 °C) and; after them, in the same year, Elbadry and Ghali developed a finite element computer program and in 1984 Cooke and Priestley did a similar study with prestressed concrete bridges and reasonable agreement was found as cited by Shushkewich [24]. In 1983 Potgieter and Gamble [21] conducted a detailed study on nonlinear thermal gradients in 26 locations of the United States and developed a finite difference computer model which does heat flow analysis and predicted gradients for that locations and they verified their results by comparing the data they get from Kishwaukee Bridge in Illinois. After that, Imbsen et al. (1985) Nutt improved Potgieter and Gamble's work and presented a comprehensive state-of-art report, NCHRP Report 276 (1985) on thermal effects in concrete bridges [19] which later formed the basis for AASHTO guide specifications regarding thermal gradients.

Prior to AASHTO 1989c [7], former AASHTO Standard Specifications for Highway Bridges considered stresses, axial expansion and contraction only because of uniform temperature changes and ignoring the probable stress distributions throughout the depth of girder. In 1989 AASHTO published their Guide Specification, Thermal Effects in Concrete Bridge Superstructures (AASHTO 1989a)

[2], which was based on NCHRP Report 276. The AASHTO Guide Specification for Design and Construction of Segmental Concrete Bridges (AASHTO 1989b) [4], required the consideration of thermal gradients in the design of all segmental bridges. The United States is divided into four solar radiation zones and a design gradient distribution across the depth of girder is specified for both positive and negative gradient, in different magnitudes for each zone. Internal stresses and structure deformations due to temperature gradients should be determined from to Table 1, according to solar radiation map of US and definition of T_1 and T_2 are explained in Figs. 1 and 2.

Table 1 Basis for temperature gradients

Zone	T_1 (°F)	T_2 (°F)
1	54	14
2	46	12
3	41	11
4	38	9

Fig. 1 Solar radiation zones for the United States, AASHTO (2012)

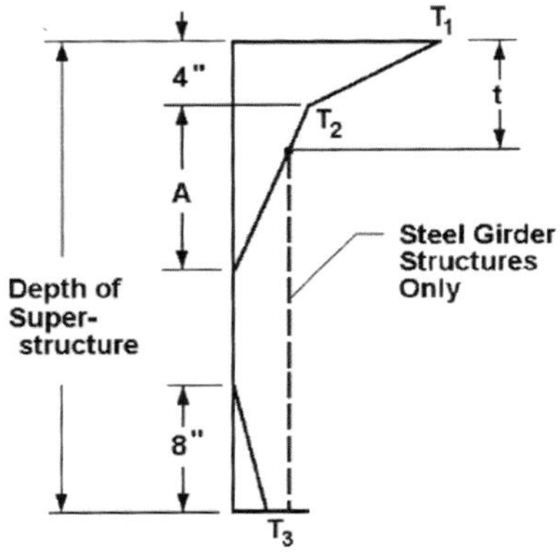

Fig. 2 Positive vertical temperature gradient in concrete and steel superstructures, AASHTO (2012)

Here dimension factor "A" is 12 in. for superstructure depths greater than 16 in. and, 4 in. less than the depth of the superstructure on the other way around and temperature T3 shall be taken as 0 °F unless a unless a site-specific study is made to determine an appropriate value, but it shall not exceed 5 °F.

After that, several field studies conducted by researchers that generally verified the positive gradient design values. The negative gradients in the AASHTO (1989a) were based on the British Standard BS 5400 (1978) [11]. The shape of negative thermal gradients has since been modified over time depending on new analysis and experimental studies. In AASHTO (1994b) [5]; the magnitude of negative gradient is reduced to both simplify the design process and reduce the high tensile stresses occurring over the top few inches of the cross section, since the amount of prestressing required for that much tension found to be quite large and resulted in requirement to place tendons that otherwise would not require none. Then the AASHTO LRFD Bridge Design Specifications (AASHTO 1998a) [1] and The AASHTO Guide Specifications for Design and Construction of Segmental Concrete Bridges (AASHTO 1999) [3] simplified the positive gradient and reduced negative gradient by 40 %. Prior to 1998, negative gradient was taken as −0.50 times the positive gradient. Also, in that codes surface conditions had been involved in design for negative gradient as −0.30 times of positive for plain concrete surfaces and −0.20 times for surfaces with 2 in. (5 cm) asphalt topping because of the insulating nature of the asphalt by reducing the loss of heat from the surface of the flange, therefore the magnitude of thermal gradient. Moreover, prior to 1994, thermal gradients were applicable only to superstructure depths greater than 2 ft but from that on, different superstructure depths are taken into

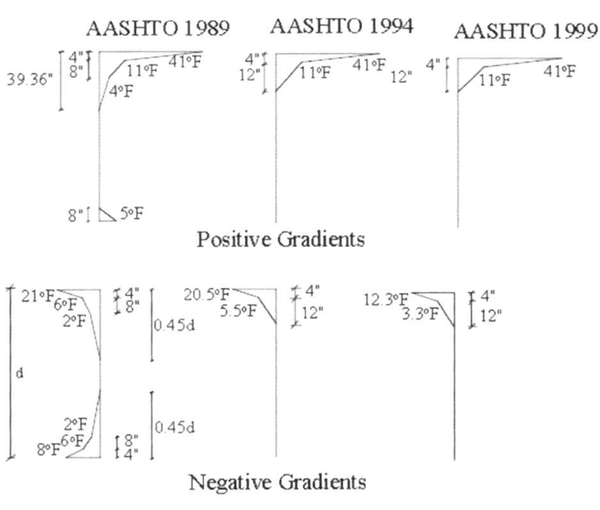

Fig. 3 Evolvement of positive and negative thermal gradients for Zone 3 (for superstructure depths greater than 2 ft, AASHTO 1989, 1994, 1999)

account in design. Nonlinear thermal gradient values and shapes are remained the identical after 1999 in other intermediary specifications and also the same in latest 2012 LRFD Specifications [6]. Here in Fig. 3, an example given for Zone 3 to the evolvement of design gradients in AASHTO through the years.

3 Formulas to Predict Temperature Gradient

Over the years, equations are developed to predict temperature gradient along the depth of the girder depending on daily air temperature, solar radiation or both.

Potgieter and Gamble developed following formula to predict temperature variation along the depth of a girder section as:

$$\Delta T_v = 28.2\left(\frac{H \cdot \alpha_s}{29.089} - 0.7\right) + 0.342(TV - 11.1) + [32.3 - 4.84\upsilon + 0.771\upsilon^2 - 0.008\upsilon^3 + 0.00463\upsilon^3] \quad (1)$$

in which TV is the air temperature variation (or the difference between the daily maximum and minimum air temperatures), and α_s is the solar absorptivity of concrete.

Roberts-Wollman et al. proposed a simpler equation which calculates the differential based on a different variable in terms of; T_{max}, maximum air temperature, T_{3-avg}, the three-day average air temperature and H, solar radiation as:

$$\Delta T_v = 0.9(T_{max} - T_{3-avg}) + 0.56(H - 12) \quad (2)$$

Later; Lee and Kalkan [16] developed the equation to calculate the maximum vertical temperature differential, ΔT_v, on a prestressed girder is:

$$\Delta T_v = \frac{3}{4}H + \frac{1}{15}(T_{max} - 2T_{min}) - \left[0.37 + 2.93\upsilon - 0.25\upsilon^2 + 0.008\upsilon^3\right] \quad (3)$$

where H is the total daily solar radiation (MJ/m^2), T_{max} and T_{min} are the daily maximum and minimum air temperatures (°C), and υ is the daily average wind speed (m/sec).

4 Research

This research is conducted as a part of a project named "Development of a Design Guideline for the Improvement of Technology in Design and Construction of Bridges in Turkey" and is coordinated by Middle East Technical University (METU) for partner organization General Direction of Highways of Turkey (KGM) and supported by The Scientific and Technological Research Council of Turkey (TÜBİTAK Project-110G093). The aim of this study is to construct a solar radiation zone map for Turkey, modelling the map of AASHTO and recommend thermal gradient values to engineers designing especially segmental box girder bridges which are much susceptible for nonlinear temperature differences along the cross section.

Comparing Aksoy's studies [8] on geographical distribution of 22-year average of annual global solar irradiation map of Turkey (Fig. 4) and Solar Radiation map of US prepared by National Renewable Energy Laboratory, Turkey is subdivided into two radiation zones similar to AASHTO Zone 1 and 2 as depicted in Fig. 5 and the calculations corresponding to thermal gradients for each zone are performed.

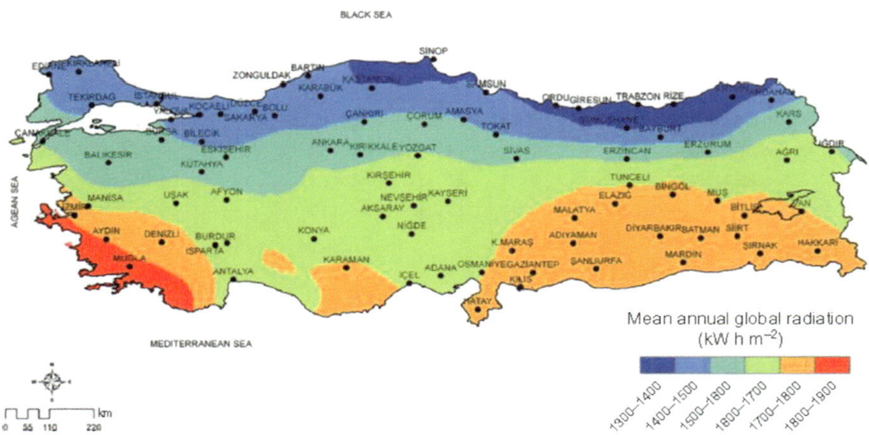

Fig. 4 Geographical distribution of 22-year average of annual global solar irradiation based on NASA/SSE (Aksoy 2011)

Fig. 5 Proposed solar radiation zones for Turkey

Eight cities from different regions of Turkey; Ankara, Antalya, Bursa, İstanbul, Izmir, Trabzon and Van are selected as representative cities of contiguous areas and the primary environmental data affecting the thermal gradient for that cities is collected (Fig. 6). Radiation data for Turkey is taken from monthly averaged solar insolation incident on a horizontal surface at 3-h intervals (in kW/m^2) from the Surface Meteorology and Solar Energy (SSE) satellite dataset of National Aeronautics and Space Administration (NASA) [18]. Hourly air temperature observation data was available in National Climatic Data Center of National Oceanic and Atmospheric Administration (NOAA) website. These data is then converted in an appropriate form in order that it can be used as input in the analysis program. In Figs. 7 and 8, a one yearly solar radiation and temperature change graph is given for a representative city.

A box girder bridge section which is identical to the one in "Precast Balanced Cantilever Bridge Design using AASHTO LRFD Bridge Design Specifications" (Fig. 9) prepared for American Segmental Bridge Institute given in Fig. 6 for span section with 5 cm asphalt topping was chosen as a representative section. This section was modelled by utilizing a two dimensional (2D) finite element heat transfer analyses program (Fig. 10): Panthalassa [9], developed in Middle East Technical

Fig. 6 Representative cities selected for the thermal gradient analysis

Fig. 7 Radiation data for Bursa between Jan. 2012 and Dec. 2012

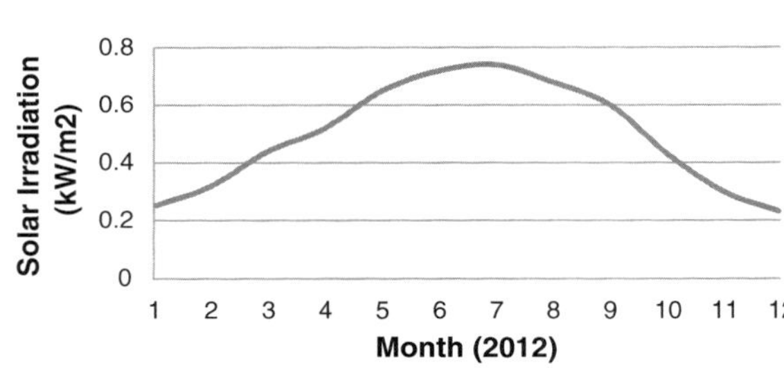

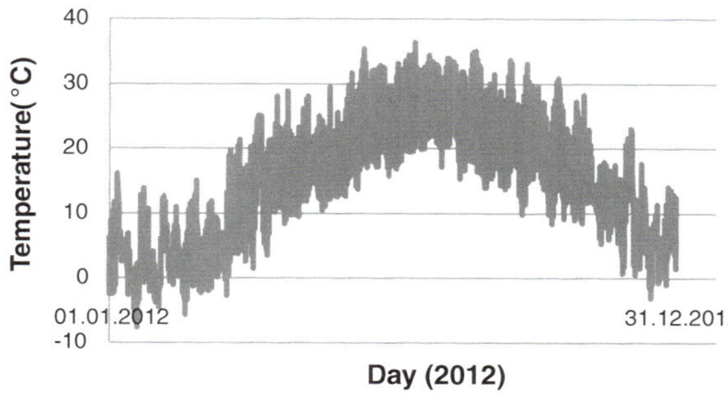

Fig. 8 Temperature data for Bursa between 01.01.2012 and 31.12.2012

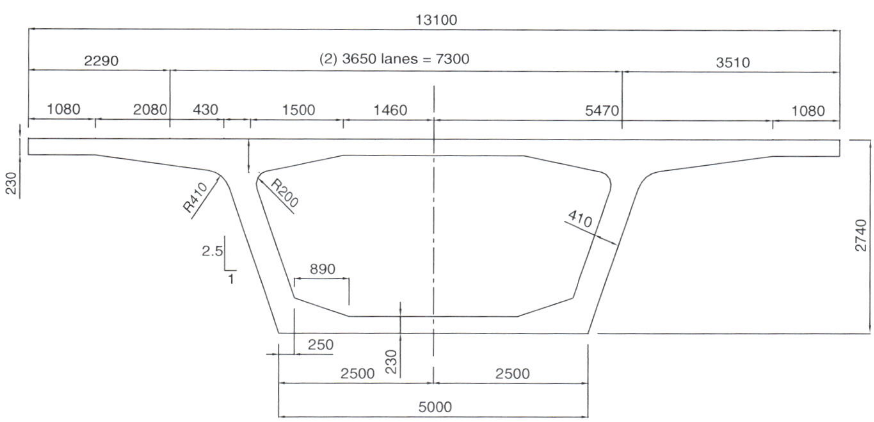

Fig. 9 Bridge girder cross section used in the analysis with its dimensions in mm

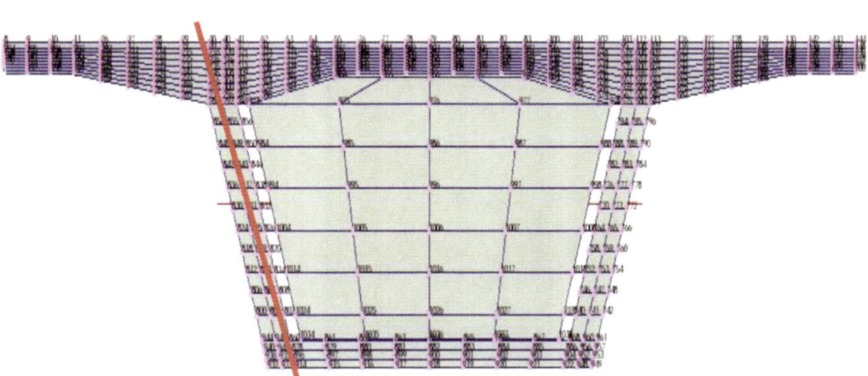

Fig. 10 Finite element model of the girder and the line along which the readings are taken

University laboratories, and sensitivity analyses were performed for the cross section. The 28 day cylinder compressive strength of concrete is taken as 35 MPa, and material properties like conductivity, film convection, specific heat and emissivity for both the concrete and asphalt are taken according to The American Society of Heating, Refrigerating and Air-Conditioning Engineers (ASHRAE) Handbook [10] and suggestions of Lee [15]. Also the air inside the box is modeled with corresponding parameters.

The solar radiation and ambient temperature data of each city obtained in year 2012 is entered to the program and after the analysis the temperature of each node in cross section is obtained to calculate the thermal gradient. Readings are obtained for a line along one of the webs in vertical direction from the topmost of the girder to the base (shown with the red line in Fig. 10). The initial temperature conditions are taken into account by starting the analysis with data of the prior month, i.e. December 2011. Results are acquired in hourly intervals in order not to omit the most critical gradient values. The highest positive gradient values tend to occur at 12 p.m. Greenwich Mean Time (2 p.m. for GMT +2 and 3 p.m. for GMT +3 time zones) in June and July while negative gradient values tend to occur at midnight times in December.

Here the five most critical positive gradient values are plotted along the depth of the girder for an example city, İstanbul (Fig. 11). In the following graph, the highest positive gradient values for each city analyzed are also plotted (Fig. 12). Generally T_3 values are so small that they can be ignored in design.

Since the T_1 gradient values actually show the temperature difference of the deck from the average of webs and T_2 as the difference between 10 cm below the top surface and the average of web, to emphasize it, they will be referred with Δ°C units hereafter:

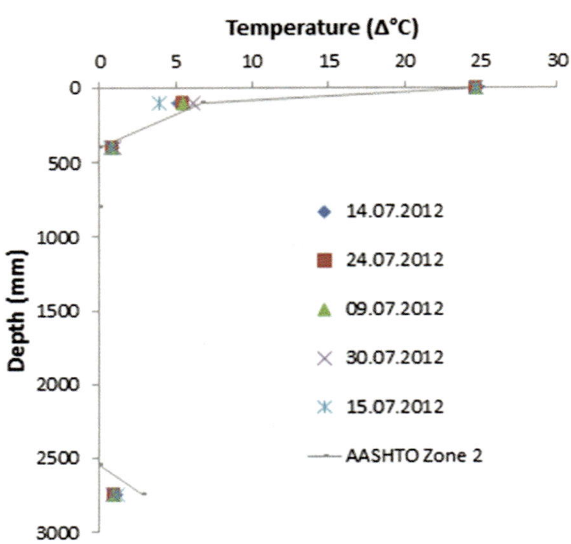

Fig. 11 Maximum measured positive gradients in İstanbul compared with recommended AASHTO Zone 2

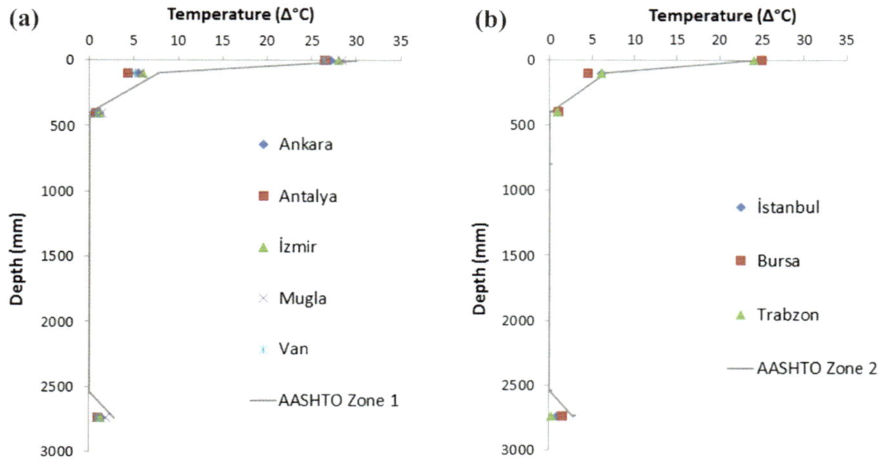

Fig. 12 Maximum measured positive gradients for 8 city analyzed for Zone 1 (**a**) and Zone 2 (**b**) compared with recommended AASHTO Zone 1 and 2

Table 2 Positive and negative temperature gradient values for the analyzed cities

Zone	City	Positive temperature gradient, Δ°C (Δ°F)		Negative temperature gradient, Δ°C (Δ°F)
		T1	T2	T1
Zone 2	İstanbul	24.7 (44.9)	6.2 (11.3)	−1.5 (−2.7)
	Bursa	24.9 (45.3)	4.4 (8.0)	−2.0 (−3.6)
	Trabzon	24.1 (43.8)	6.1 (11.1)	−1.0 (−1.8)
Zone 1	İzmir	28 (50.9)	5.9 (10.7)	−0.9 (−1.6)
	Muğla	28.5 (51.8)	5.7 (10.4)	−1.1 (−2.0)
	Ankara	27.1 (49.3)	5.5 (10)	−2.4 (−4.4)
	Antalya	26.4 (48.0)	4.3 (7.8)	−0.9 (−1.6)
	Van	26.3 (47.8)	5.1 (9.3)	−2.0 (−3.6)

In the following Table 2, the analyses results showing the maximum temperature differential between top of concrete girder and average temperature of web for each city are given.

Per the given analysis results; a 24.6 °C temperature difference for Zone 2 and 27.3 °C for Zone 1 is obtained for T_1 as the averages of the corresponding analyzed locations. The value for Zone 2 is the same as the AASHTO recommendation, and the value for Zone 1 is relatively below but very close to the AASHTO value. Therefore; it is suggested to use Δ 25 °C for Zone 2 and Δ 28 °C for Zone 1 for design of bridges for thermal gradient. The temperature difference 10 cm below the top surface, namely ΔT_2 values can be taken as 6 °C for both of the zones. Since the negative gradient values after the analysis is obtained are quite smaller than AASHTO recommendations; it will be convenient to use −0.30 times of the positive gradient value for the bridges which have decks of which top surfaces has asphalt topping less 5 cm, and −0.20 times of positive gradient for the decks which have more insulating materials.

5 Conclusions

The solar radiation data obtained for different parts of Turkey is used to determine the thermal gradients at the depth of the segmental concrete box bridges. The extensive analysis of the results indicate that Turkey can be divided into two solar zones and the corresponding thermal gradient values are in good agreement with slight differences in values for the ones suggested for the bridges in US soils. The differences in results are less than five percent. It can be concluded that the findings of this study can be applied to be used in thermal gradient analysis of segmental concrete box bridges in Turkey.

References

1. AASHTO (1998a) AASHTO LRFD bridge design specifications. Washington, D.C
2. AASHTO (1989a) Guide specifications for design and construction of segmental bridges. American Association of State Highway and Transportation Officials, Washington, D.C
3. AASHTO (1999) Guide specifications for design and construction of segmental bridges. American Association of State Highway and Transportation Officials, Washington, D.C
4. AASHTO (1989b) Guide specifications: thermal effects in concrete bridge superstructures. American Association of State Highway and Transportation Officials, Washington, D.C
5. AASHTO (1994b) Interim specifications for the guide specifications for design and construction of segmental bridges. American Association of State Highway and Transportation Officials, Washington, D.C
6. AASHTO (2012) LRFD bridge design specifications. Customary U.S. Units, Washington, D.C
7. AASHTO (1989c) Standard specifications for highway bridges. American Association of State Highway and Transportation Officials, Washington, D.C
8. Aksoy B (2011) Solar radiation over Turkey and its analysis. Int J Remote Sens
9. Albostan U (2013) Implementation of coupled thermal and structural analysis methods for reinforced concrete structures. Middle East Technical University
10. ASHRAE (2001) The American Society of Heating, Refrigerating and Air-Conditioning Engineers Handbook
11. British Standard BS 5400 (1978) Steel, concrete and composite bridges, part I, general statement. British Standards Institution, Crowthorne, Berkshire
12. Deutsche Norman DIN 1072 (1967) Strassen and WegbrUcken (Street and road bridges), II Berl in and Koln
13. Hoffman PC, McClure RM, West HH (1980) Temperature studies for an experimental bridge, research project 75-3 Interim report. Pennsylvania State University
14. Imbsen RA, Vandershof DE, Schamber RA, Nutt RV (1985) Thermal effects in concrete bridge superstructures, NCHRP 276. Transportation Research Board, Washington D.C

15. Lee JH (2012) Investigation of extreme environmental conditions and design thermal gradients during construction for prestressed concrete bridge girders. J Bridge Eng ASCE
16. Lee JH, Kalkan I (2012) Analysis of thermal environmental effects on precast, prestressed concrete bridge girders: temperature differentials and thermal deformations. Adv Struct Eng 15(3)
17. Leonhardt F, Kolbe G, Peter J (1965) Temperaturunterschiede gefahrden spannbetonbrilcke (Temperature differences endanger prestressed concrete bridges), 11 Beton und Stahlbetonbau 60(7):231–244
18. National Aeronautics And Space Administration (NASA) (2014) Surface meteorology and solar energy. Available online at http://eosweb.larc.nasa.gov/
19. National Renewable Energy Laboratory (NREL) (1994) Solar radiation data manual for flatplate and concentrating collectors, Golden, CO
20. Post-Tensioning Institute (PTI) (1997) Precast segmental box girder bridge manual. Post-Tensioning Institute and Precast/Prestressed Concrete Institute
21. Potgieter IC, Gamble WL (1983) Response of highway bridges to nonlinear temperature distributions. Structural research series no. 505. University of Illinois at Urbana-Champaign, Illinois
22. Priestley MJN (1987) Design of concrete bridges for temperature gradients. ACI J 75(5):209–217
23. Roberts CL (1993) Measurement based revisions for segmental bridge design and construction criteria. The University of Texas at Austin, Austin
24. Shushkewich KW (1998) Design of segmental bridges for thermal gradient. PCI J 43(4):120–137
25. Wood BA (2007) Thermal gradients and their effects on segmental concrete box girder bridges. University of Texas at Austin

Part II
Construction and Erection Techniques

Risk Assessment and Management for Bridges

Xin Ruan

Abstract Risk assessment and management has gained more and more popularity in bridge engineering due to its direction on preventing or alleviating unfavorable situations during the life cycle of bridges. In the current applications, risk management is recognized as an extension of risk assessment to provide management strategies. In fact, risk management is essential to the improvement of decision efficiency and validity since it is the direct procedure connecting risk assessment and on-site users. To enhance the influence of risk management, management oriented risk assessment is proposed. Introductions to the basic conceptions of risk assessment and management are presented. Detailed instructions on the proposed approach are provided with consideration to different management requirements and stages. Four applications of the management oriented assessment are introduced in detail on how the approach is realized in practical situations.

Keywords Risk management · Risk assessment · Bridge engineering

1 Introduction

The word "risk" has a long history starting as the "risqué", while the practice of risk analysis and management were performed in 3200 B.C. by Asipu [1]. With a vast time of development, modern risk analysis didn't appear until last century. In 1992, value-at-risk was firstly introduced by New York Stock Exchange (NYSE) for informal capital tests on its member firms. Since then, risk analysis has become an extensively adopted decision assisting tool in the finance fields, and gained its popularity in other industries as well in recent years.

X. Ruan (✉)
Department of Bridge Engineering, Tongji University, 1239, Siping Road, Shanghai, China
e-mail: ruanxin@tongji.edu.cn

A. Caner et al. (eds.), *Developments in International Bridge Engineering*,
Springer Tracts on Transportation and Traffic 9,
DOI 10.1007/978-3-319-19785-2_8

Risk analysis and assessment is applied widely in bridge engineering due to the fact that there are many uncertainties needed to be analyzed and management in the whole life cycle of bridges. Usually, bridges are of significant importance to regional transportation and communication. It is imperative to guarantee the performance from the design stage to the operation stage of bridges. To realize this goal, bridge engineers have to manage the enormous possible risk events brought by the uncertainties. Recent development of risk assessment in bridge engineering has offered the related bridge managers with promising prospects in controlling the risks and their unfavorable consequences.

In spite of the extensive applications of risk assessment and management for bridges, there are still problems that need further improvements. One most obvious deficiency is that the conception of risk management is not as valued as it is supposed to be. In current practice, risk management is generally regarded as a subsequent procedure based on the results of risk assessment, while it should be recognized that risk management is of great importance since it is the direct presentation from the risk assessors to the on-site users. Instead of making the risk management merely based on risk assessment results, the risk assessment should be performed with consideration to the requirements of risk management.

In order to solve the contradiction between risk management and assessment for bridges, management oriented assessment conception is proposed in this paper. Firstly, an introduction to popular risk assessment and management procedures in current applications is presented. Different stages of risk assessment and characteristics of engineering risk management are introduced. Then management oriented risk assessment for bridges is elaborated. The purpose of the risk management is studied and adopted as a determining factor for the selection of risk assessment methods. Specifications on how to perform management oriented risk assessment in different stages of bridges are explained in details. Finally, four engineering application cases are presented. In each case, the parts demonstrating the conception of management oriented assessment are highlighted.

2 Risk Assessment and Management in Bridge Engineering

To perform risk assessment and management, a definition of the project risk is necessary to clarify the subject to be assessed and managed. However, it is extremely difficult to propose a universal definition of risk. Since its foundation, Society of Risk Analysis tried to define "risk" but ended up with giving up after four-year endeavor. Finally, it held that different definitions of risk can be obtained by different assessors as long as specific explanations are provided [2]. In the applications in bridge engineering, definition of risk can be varied according to different management targets.

Risk assessment is the imperative prerequisite of risk communication and management. In the current practice, there are several risk assessment approaches provided for bridge engineering applications, such as Preliminary Hazard Analysis Method, Failure Mode and Effect Analysis Method, Event Tree Analysis Method and Risk Matrix Method. As a systematic process for identifying risk sources and quantifying and describing the nature, likelihood, and magnitude of risk associated with some situation, action, or event that includes consideration of relevant uncertainties, risk assessment generally consists of three major stages, (1) risk identification, (2) risk analysis, and (3) risk evaluation.

Risk identification is a discrimination process on realistic and potential risk characteristics by means of perception, judgment and classification. As the preliminary evaluation on the project risks, risk identification is the foundation of risk analysis and evaluation. The results of risk identification process provide the following two stages with study objects, which are supposed to cover all the possible scenarios. Hence, the correctness and comprehensiveness of risk identification are substantially important to the whole assessment process. For the risk identification, several methods are usually adopted in engineering application, such as brainstorming, cause and effect diagram method, flow chart method and Delphi method.

Risk analysis is a determining process of assessing the likelihood of occurrence and potential consequences of risk scenarios identified [3]. Risk analysis can be performed by qualitative or quantitative ways, determined by the features of the risk scenarios. The analysis results on the probabilities and consequences are demonstrated by numbers, rankings or graphic expressions, in accordance with the selected risk evaluation method. For the practice in bridge engineering, risk analysis is complicated since the scenarios can be significantly varied. Determination of the probability and consequence of a risk scenario can be very vague since it usually involves with many different factors such as personnel safety, structure damage and economic loss.

Risk evaluation is an assigning process in which all the identified and analyzed scenarios are designated with its corresponding risk level by a selected evaluation method. By indicating the risk level of each scenario, risk evaluation process provides the risk management with informative results obtained from risk identification and analysis, connecting the risk assessment and management. Risk evaluation is the essence of the risk assessment, and usually the risk assessment method refers to the risk evaluation method. The selection of the risk evaluation method should consider the management purpose and actual situation of the project. Once the evaluation method is determined, the form of the probability and consequence of risk scenarios is certain, with which the form adopted in risk analysis process should be in accordance.

The risk assessment process not only provides the assessors with a comprehensive view on the risk scenarios identified, but also obtains the initial conditions for risk management. Risk management is the application of a systematic process for identifying, analyzing, planning, monitoring, and responding to risk so that critical infrastructure will meet service, safety, health, and welfare expectations. With the assistance of the risk assessment presenting the risk levels of scenarios to be managed, risk management can be implemented with confidence in actually improving the decision efficiency.

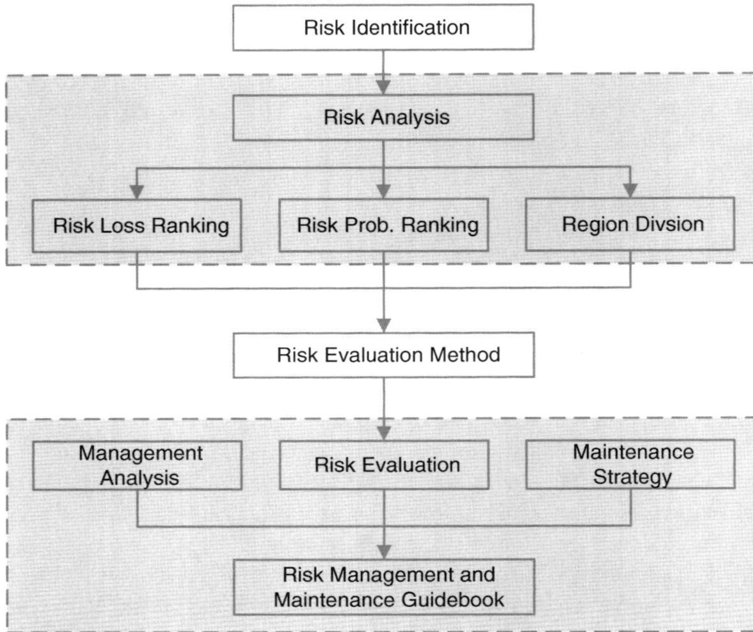

Fig. 1 A typical risk assessment and management process for bridges during operation

Risk management is usually established on the risk assessment results only, which indicates that risk assessment is performed without the influence of risk management. For applications in bridge engineering, strategies formulated during risk management process are directly presented to the sponsors. In some cases, the strategies are used by on-site staff with little engineering experience. Under these circumstances, risk management should be emphasized on its simplicity and specificity, which can be easily ignored in the risk assessment process. A typical risk assessment and management process is provided in Fig. 1.

To obtain more satisfying outcomes, the risk assessment should take the requirements of a simplified and specific risk management process into consideration, which means risk assessment should be oriented by management. In this way, the target of risk assessment is set to serve for the benefits of risk management. During the assessment, conception of the project risk can be clarified, knowing the specific requirements of the management.

3 Management Oriented Risk Assessment for Bridges

To perform the risk assessment under the orientation of management requirements, features of risk management for bridges should be studied thoroughly. For each important feature, corresponding measurement should be adopted in the risk assessment. Bridge is a structural complex consisting of significantly different components with various characteristics. According to the management scope, the assessment targets in bridge engineering can be separate component with specific management goals, or the whole bridge with overall management requirements.

For the risk management of specific components, risk scenarios to be assessed and managed are certain in most cases. Risk identification is much simpler under this circumstance, since the screening process is no longer necessary. Risk analysis can be directly implemented on the risk scenarios under consideration. The method adopted to analyze the probability and consequences of the scenarios can be diverse according to the specific objects the management process supposed to deal with. The corresponding evaluation criteria is determined by the adopted analysis method. Compared with traditional analysis method, there are several common analysis methods in wide adoption, such as life cycle cost analysis and reliability analysis.

When the risk management is performed on the integral level of a bridge, the overall performance of the bridge is much more complicated than the situation of considering only one single type of the component. All the risk scenarios should be taken into consideration to obtain a more comprehensive assessment and to avoid possible failure due to ignored risks. The typical three stages of risk assessment are all imperative under this circumstance. The characteristics of the risk scenarios are essential and diverse due to the complicated situation. During the risk identification, all the characteristics should be studied to acquire a well-founded database for further research. Risk analysis and evaluation are comparatively confined to the compatible procedures that satisfy various requirements of different types of risk scenarios.

Under both specific and general management situations, risk analysis is always a key step to the correctness and validity of the whole risk assessment. The risk analysis approach is essential to that and should be determined majorly by the management level of the project, characteristics of risk scenarios, structure features and financial situation. When the condition of the project approves, component vulnerability and structure robustness analysis are necessary to obtain more detailed and comprehensive understanding on the risk scenarios.

Component vulnerability is defined as the damage probability of a component when exposed to potential risk sources. The purpose of component vulnerability analysis is providing the risk analysts with an explicit conception of the damage condition of the component in consideration under certain risk scenarios. Structure robustness is defined as the damage sensitivity of the whole structure caused by failures of components. Structure robustness analysis is generally employed to determine the safety conditions of the whole structure giving that one or multi certain components fail. Figure 2 demonstrates a recommended vulnerability and robustness analysis process for the stayed cables of a cable-stayed bridge under earthquake scenario [4].

Risk management can be very different when performed in different stages of the life cycle of bridges. In the design stage, the practical experience of the design staff is the most important aspect of the risk assessment and management. With the assistance of the design code, bridge designers are supposed to use their

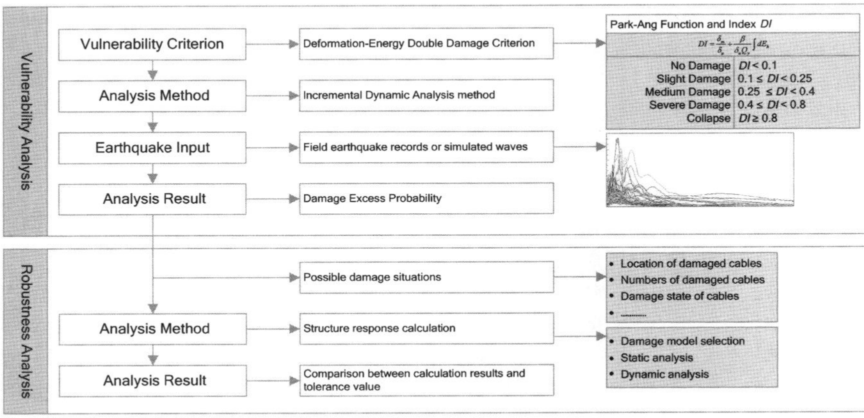

Fig. 2 Vulnerability and robustness analysis process of stayed cables under earthquake scenario

expertise to avoid construction and operation risks as much as possible. It has to be realized that risks are inevitable in the whole life cycle of bridges. The risk assessment and management during design stage should focus on the effort to prevent major design mistakes, such as the location selection, material adoption, design of concrete cover, structural types and so on.

As for the construction stage, bridges are usually structurally unstable, making the structure the most vulnerable time in the whole life cycle. Safety issues should be emphasized during the construction, and accordingly are the focus of the risk management. The safety consists of two parts, personnel safety and structural safety. During the risk analysis, structural safety can be analyzed in detail by referring to component vulnerability and structure robustness analysis. It is highly recommended to introduce these analysis when the bridge structure is complicated or the intermediate forms of the structure are unstable. Also, construction schedule is greatly valued due to consideration on money-saving and administration requirements. During the risk evaluation, more weights should be assigned to the personnel and structure safety and delayed time to emphasize on their substantial influence to the management.

Currently, most risk management and assessment projects are designed for bridges in operation due to more complicated situations on bridges. When the risk management is performed during operation stage, it is common that the formulated strategies are carried out directly by the on-site staff, whose expertise knowledge in bridge engineering may be not quite complete. This situation demands that the output of risk assessment should be delivered in specific and apprehensible ways. Also, a clear graphic demonstration on the risk scenarios to be managed is very helpful to assist the on-site staff with making prompt responses. Risk scenarios identified can be categorized by the regions in which they are expected to appear, providing the risk identification and analysis with convenient indications.

4 Case Study

4.1 Life Cycle Cost Analysis for Wind Barrier of Hangzhou Bay Bridge

Hangzhou Bay Bridge is a cross-sea bridge with a total length of approximately 36 km in Zhejiang province, China. Crosswind becomes an important factor that affects vehicle travel safety due to the deck altitude and complex environment of the sea. To control the risk of vehicle rollover and other unfavorable consequences brought by the crossing wind, a study aiming at proposing an optimized wind barrier scheme is presented [5].

The problem in this risk management project is specific, providing only one risk scenario. Analysis on this scenario is performed by introducing a direct approach, i.e. life cycle cost analysis. Several forms of wind barrier are proposed including local wind shielding screen around tower, wind shielding screen with height of 1.3 m. Based on wind environment study result, the whole bridge are divided into several sections, and different kinds of countermeasure are proposed for each section. Finally, five schemes are submitted for discussion and further study. Figure 3 gives a vivid illuminate of each scheme.

To compare the proposed five schemes of wind barrier, a cost-benefit analysis is performed. The cost calculation is simply related to the construction and replacement fee of the wind barriers in each scheme. The benefit calculation is more complicated since it involves the weather condition at site and the utility of each scheme on reducing the wind speed. The benefits from installing the wind barrier consist of two parts, the saved detour expense for vehicle drivers and the improved income from keeping the bridge open during strong wind for the managers. The cost-benefit analysis results of both parts are provided in Tables 1 and 2.

According to the results, it is indicated that scheme B has the shortest cost-benefit equivalent time for both parties and the total benefit of 25 years are acceptable, making it the most economic management scheme among all the five proposed solutions.

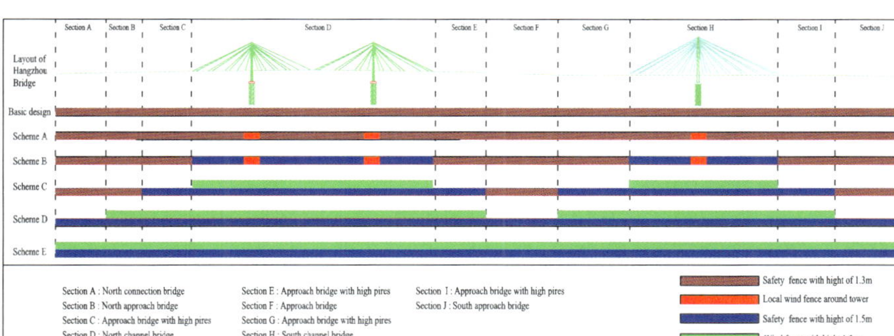

Fig. 3 Wind barrier schemes of Hangzhou Bay Bridge

Table 1 Cost-benefit analysis result for vehicle drivers (in Million RMB)

	Additional cost	Benefit of each year	Total benefit of 25 years	Cost-benefit equivalent time (year)
Scheme A	1.78	17.11	513.3	0.10
Scheme B	2.31	28.11	843.3	0.08
Scheme C	10.57	43.39	1301.7	0.24
Scheme D	71.35	49.32	1479.6	1.45
Scheme E	149.29	53.79	1613.7	2.78

Table 2 Cost-benefit analysis result for bridge managers (in Million RMB)

	Additional cost	Benefit of each year	Total benefit of 25 years	Cost-benefit equivalent time (year)
Scheme A	1.78	3.35	83.87	0.53
Scheme B	2.31	5.51	137.83	0.42
Scheme C	10.57	8.51	212.82	1.24
Scheme D	71.35	9.68	241.92	7.37
Scheme E	149.29	10.55	263.86	14.14

4.2 Vulnerability Analysis of Minpu Bridge Under Earthquake

Minpu Bridge is located in Shanghai, connecting both sides of the city. Minpu Bridge is the first long-span double-layer steel cable-stayed bridge in China, with a 708-m main span. Double cable plane is designed with 88 stayed cables on each side and in total symmetric arrangement. The upper layer is designed as expressway with 8 lanes and a width of 40.5 m, while the lower layer is designed for local traffic with 6 lanes and a width of 26.0 m.

Risk management for Minpu Bridge during operation is required to be explicit by the project managers [6]. For the essential risk scenarios to the bridge, structure safety is demanded to be sufficient. After the preliminary risk identification, it can be inferred that the earthquake is the major possible reason for the structure failure. Based on this judgment, a component vulnerability analysis is presented under the influence of earthquake. Considering the imperative functions of pylon, whose failure is totally unacceptable, the component vulnerability analysis is conducted for the pylon.

To evaluate damage level of the pylon caused by earthquake, a damage model, expressing the potential damage of reinforced concrete components as a function of the maximum deformation and the absorbed hysteretic energy, is adopted. To realize the vulnerability analysis, the Incremental Dynamic Analysis method is chosen and 59 earthquake records are calculated for a probability assessment.

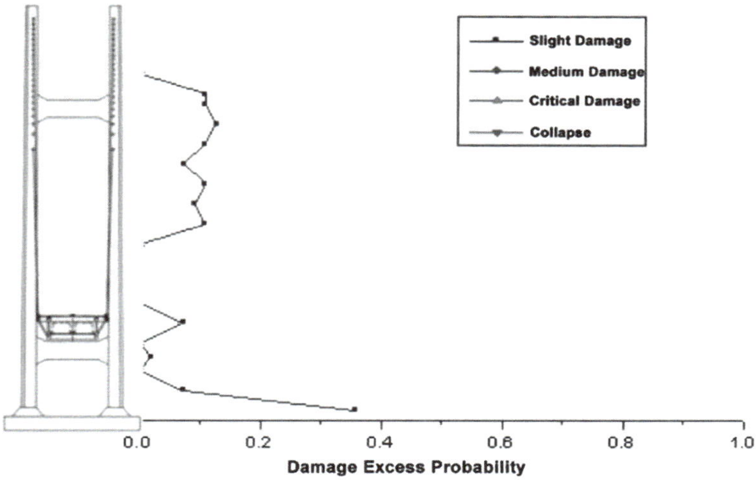

Fig. 4 Damage excess probability of the pylon when PGA = 0.4 g

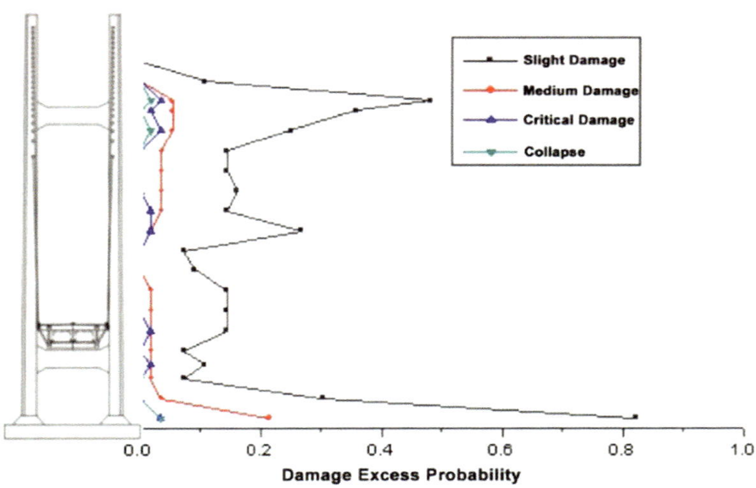

Fig. 5 Damage excess probability of the pylon when PGA = 0.6 g

Analysis results, as are shown in Figs. 4 and 5, indicate that no obvious damage will appear under earthquakes with a seismic intensity no bigger than 7°, and slight damage with a low probability in the bottom of the tower will occur under earthquake with a seismic intensity of 8°.

According to earthquake vulnerability analysis results, management oriented suggestion are presented as follows:

1. No particular countermeasure is needed under weak earthquakes;
2. The bottom part of the pylon needs to be inspected carefully after medium earthquakes and maintained when necessary;
3. Repair the damages in the bottom and cross beam parts after strong earthquakes, cease the traffic until a safety assessment on the pylon is performed and the operation safety is guaranteed.

4.3 *Typical Considerations on Risks During Construction Period of Sutong Bridge*

Sutong Bridge is a cable-stayed bridge with a main span of 1088 m, which was the record span of its kind of bridges in the world. The bridge is located in the estuary of the Yangzi River in China, thus the environment for the construction is highly tough, especially for the construction of the 306 m height pylon. Construction of the high pylon of Sutong Bridge is complicated and is influenced by different factors [7]. A brief illustration of the pylon construction process is provided in Fig. 6.

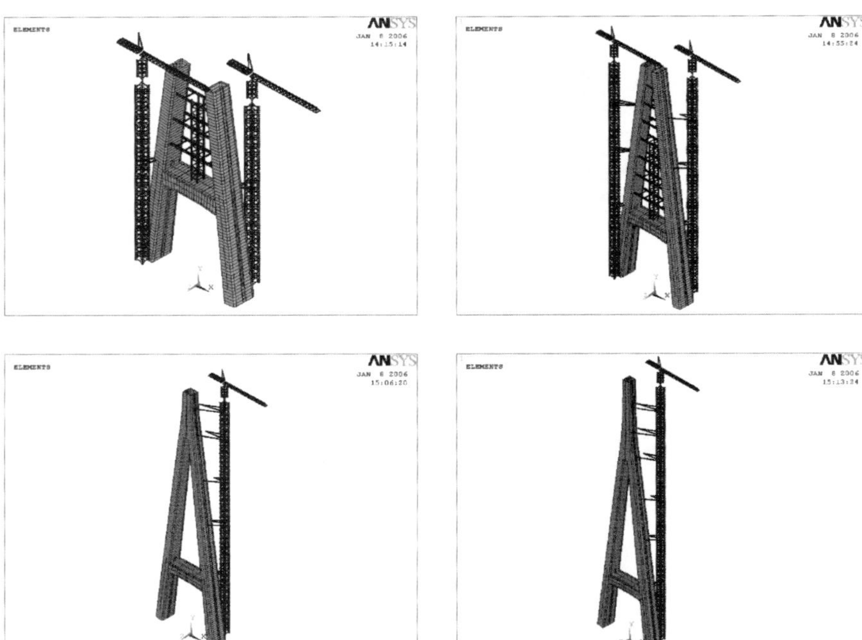

Fig. 6 Construction process of the pylon of Sutong Bridge

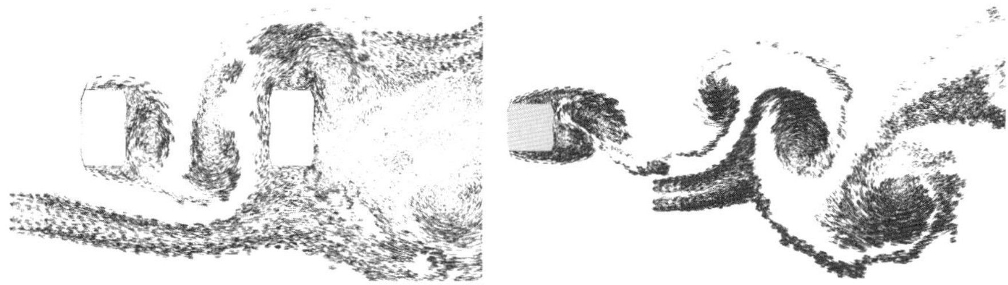

Fig. 7 Flue filed map of pylon and cranes calculated by CFD

For the risk management during construction, it is very important to analyze the following three aspects, construction schedule (C.S.), worker safety (W.S.) and investment increasing (I.I.), when the project loss is considered. To obtain the status of the three aspects, vortex excited resonance of the cranes which may cause significant delay and loss during the pylon construction is studied in detail.

The vortex excited resonance is a kind of self-excited limit-amplitude vibration due to wind. Discrete vortex method based CFD calculations are carried out to every key construction stages, and the lock-in wind speeds in different yaw angles of wind are obtained. After the CFD calculation, it is realized that the vortex excited resonances of pylon and cranes are likely to happen during the construction stage. However the lock-in wind speeds are limited in a low level. Considering the material damping of pylons, which are made of concrete, is large enough to absorb the energy due to vortex excited resonance of the pylon and cranes system (Fig. 7).

According to the above analysis, expects' suggestions and literature study, the analysis results for the C.S., W.S. and I.I. of 31 identified risk scenarios during the pylon construction of Sutong Bridge are obtained and listed in Table 3.

4.4 Improved Risk Matrix Based on the Risk Attitudes of Decision Makers

Risk attitudes of decision makers are significantly different in different projects. To make the management work more practical, it is important to present the risk attitudes of the decision makers during the assessment process. Usually in a risk assessment and management project, the decision wills of the managers are of the most essence. To present a management oriented risk assessment, the risk attitudes of the managers must be reflected during the assessment.

Risk attitudes can be reflected in both risk analysis and evaluation stages, yet the more direct and unbiased reflection would be in the evaluation method. Currently, risk matrix method is a widely adopted approach to perform risk evaluation in bridge engineering. Unfortunately, it has been recognized that there are

Table 3 Risk loss of the pylon construction process

No.	Description of the risk event	C.S.	W. S.	I. I.
1	Influence of earthquake to the pylon construction system	2	3	3
2	Influence of traffic ship impaction to pylon construction system	2	1	3
3	Influence of workboat impaction to the pylon construction system	1	1	3
4	Cranes vortex excited resonance	3	3	3
…	…	…	…	…
30	Insufficient density of the concrete under the anchor area	3	1	3
31	Controlling of the tensioning force to the transverse bracing during the construction of the mid-leg	2	1	3

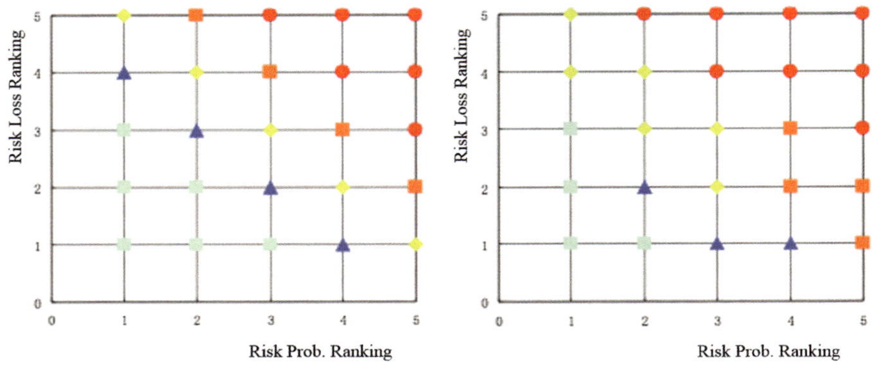

Fig. 8 Investigation results of the risk attitudes of decision makers

limitations in the applications of risk matrix method, one of which is the popular adoption of the "standard risk matrix" [8]. This limitation causes a significantly unfavorable result that all the evaluation process are based on one consistent criteria. The actual decision wills of the managers are not reflected during the evaluation process.

In order to improve this situation, attempts have been made on introducing the risk attitudes to risk matrix. A proper approach is proposed in the application of risk assessment for Changzhou Viaduct. Questionnaires are handed out to project managers regarding their opinions on the possible loss on site, from which the risk attitudes of the decision makers are studied. Demonstration of their choices under uncertain situations are presented in Fig. 8. In the figure, every dot presents a certain risk ranking at a point with certain risk probability and loss rankings. The dots in the figure from upright to bottom left represent five risk rankings from "Unacceptable", "Strict Control", "Reasonable Control", "Acceptable" to "Neglectable".

Based on the investigation results, an improvement on the risk matrix can be obtained by considering the choices of decision makers under uncertainties.

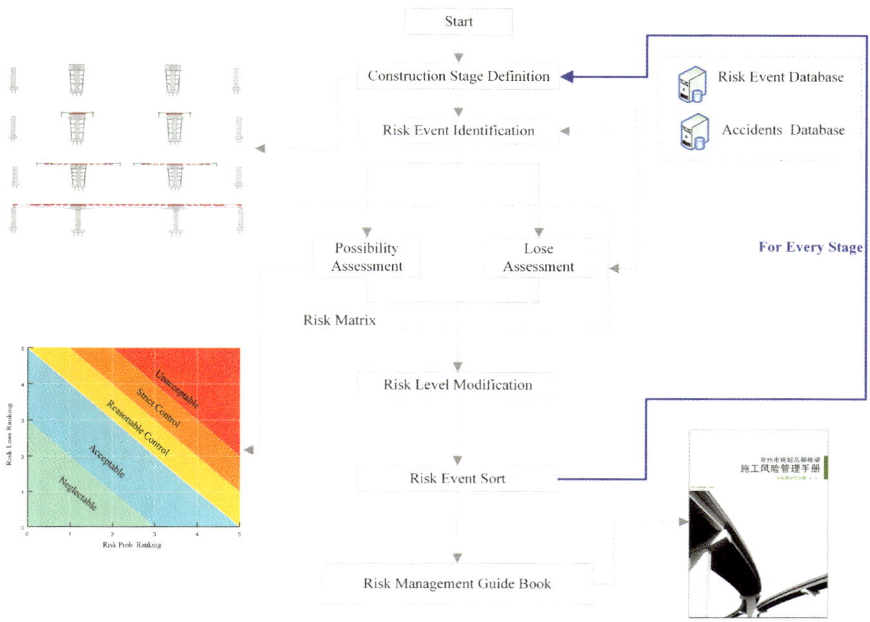

Fig. 9 Risk assessment method based on improved risk matrix with consideration on risk attitudes of decision makers

Instead of the fixed rectangular areas in traditional risk matrices, diagonal areas are adopted for the evaluation. A detailed improved risk assessment process on the base of the new risk matrix is presented in Fig. 9, in which the new matrix is illustrated as well. With this improved approach, risk evaluation results will be more practical and valid to the project managers.

5 Conclusion

Management oriented risk assessment is a very important conception in the applications of bridge engineering. Currently, related practice of this conception has been proved to be valid and efficient. Risk assessment can be more precise and directive to the actual requirements of the management. Risk communication based on this conception is extensive, and the results of communication can be expressed more thoroughly in the management strategies formulation. With all stages in the bridge life cycle covered, this conception provides the risk assessment and management with a more comprehensive and instructive sense in practice. In the future research, an integral theory on the management oriented assessment is expected to be studied and proposed.

References

1. Covello VT, Mumpower J (1985) Risk analysis and risk management: an historical perspective. Risk Anal 5(2):103–120
2. Kaplan S (1997) The words of risk analysis. Risk Anal 17(4):407–417
3. Kirchsteiger C (1999) On the use of probabilistic and deterministic methods in risk analysis. J Loss Prev Process Ind 12(5):399–419
4. Ruan X, Yin Z, Chen A (2013) Risk management strategy for long span bridge during operation stage. In: Conference on structural engineering in hazard mitigation, 2013. Institution of Structural Engineers
5. Ruan X, Chen A, Wang D (2006) Risk analysis and decision making method of wind shielding screen for Hangzhou Bay Bridge. In: IABSE symposium report, 2006, vol 1. International Association for Bridge and Structural Engineering, pp 102–106
6. Ruan X, Yin Z, Yan Z (2012) Risk based management of Minpu Bridge. In: 6th international conference on bridge maintenance, safety and management, 2012. IABMAS, pp 2638–2643
7. Chen A, Ruan X, Ouyang X-Y, Zhang J (2007) Risk management system for pylon construction of a long span cable-stayed bridge. In: IABSE symposium report, 2007, vol 4. International Association for Bridge and Structural Engineering, pp 23–30
8. Ruan X, Yin Z, Chen A (2013) A review on risk matrix method and its engineering application. J Tongji Univ 41(3):381–385

Construction of the Nhat Tan Bridge Superstructure

M. Nakayama, T. Tokuchi, H. Kawashima, S. Kaifuku, K. Matsuno and T. Nishi

Abstract The main bridge of Nhat Tan Project is a six span continuous cable stayed bridge under construction in Hanoi, Vietnam. On completion, the 1500 m bridge will be the longest cable stayed bridge in Vietnam, with 4 main spans and 2 side spans of lengths 300 and 150 m respectively. The bridge will have 8 traffic lanes and has a total deck width of 35.6 m. The superstructure of the bridge consists of a composite girder and New Parallel Wire Strands (New PWS) stay cables with a total of over 14,500 metric tons of steel, 37,500 metric tons of reinforced concrete deck slab and 1800 (220 Nos.) metric tons of prefabricated stay cables. This paper details the limitations faced in development of a construction method to ensure the safe, efficient and expeditious construction of the superstructure. In addition, it describes the innovations made to overcome challenges that arose during the implementation of the construction method.

1 Introduction

The Nhat Tan Project is constructed for a part of Ring road No. 3 located on northern side of Hanoi city, Vietnam and across the Hong River. The bridge is expected to improve the traffic access between city center and Noibai International airport [1], Northern industrial zone and some highways for connecting with other cities (Fig. 1).

IHI Infrastructure system Co., Ltd. (IIS) is performing the superstructure works of total length 3080 m including six spans continuous composite girder cable stayed bridge in the Nhat Tan Bridge Construction Project Contract Package 1. Especially this paper describes the construction method to ensure the safe,

M. Nakayama · T. Tokuchi (✉) · H. Kawashima · S. Kaifuku · K. Matsuno · T. Nishi
Department of IHI Infrastructure Systems Co., Ltd, 3-banchi, Ohama-nishimachi, Sakai-ku, Sakai-city, Osaka 590-0977, Japan
e-mail: tomonobu_tokuchi@ihi.co.jp

A. Caner et al. (eds.), *Developments in International Bridge Engineering*, Springer Tracts on Transportation and Traffic 9,
DOI 10.1007/978-3-319-19785-2_9

Fig. 1 All view of bridge after closing

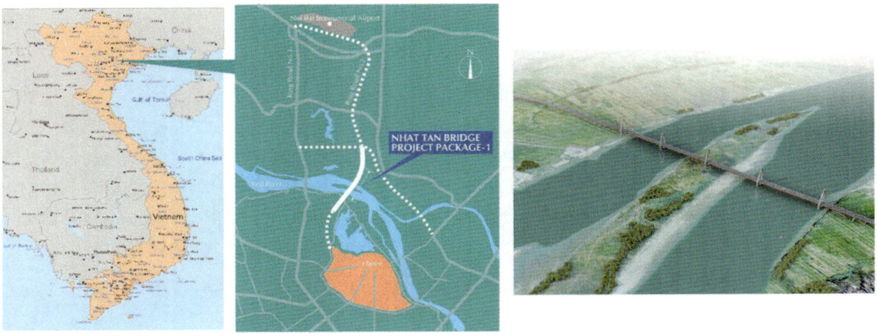

Fig. 2 Location map of the Nhat Tan Bridge project and CG image on site

efficient and expeditious construction of six spans continuous composite girder cable stayed bridge for the location shown above (Fig. 2).

The Main Bridge is six spans continuous composite girder cable stayed bridge with length of 1500 m (150 m + 4@300 m + 150 m) and effective width of 33.5 m. Five A-shaped RC pylons (H = 111 m) connected 220 numbers of stay cable support 35.6 m reinforced concrete deck slab on the 2-I-beam steel edge girder and I-beam type of steel floor beam which space 4 m. Steel anchor boxes are installed on pylon structure for anchoring stay cables.

Girder and anchor box with a total of over 14,500 metric tons are built for main structure painted by Red color which be imaged the National flag of Vietnam. They might be high impression for around the bridge location.

Main structure of steel girder is combined Edge girder for longitudinal direction and Floor beam for transverse direction (Figs. 3 and 4).

Stay cable is New Parallel Wire Strands with Ø7 mm galvanized high strength steel wire covered with High Density Polyethylene (HDPE) shaped indent pattern on the surface.

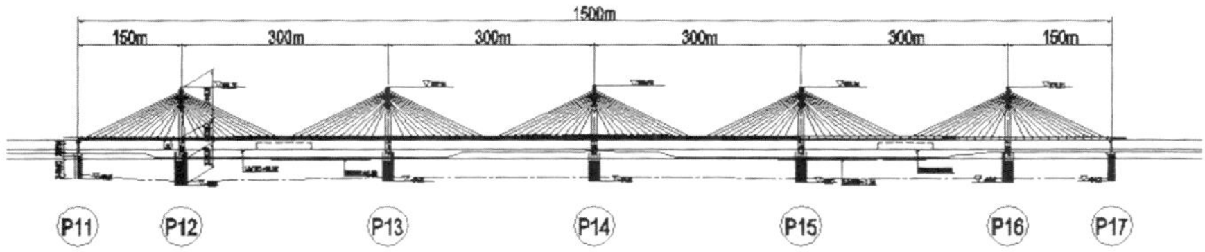

Fig. 3 General profile of main bridge of the Nhat Tan Bridge project

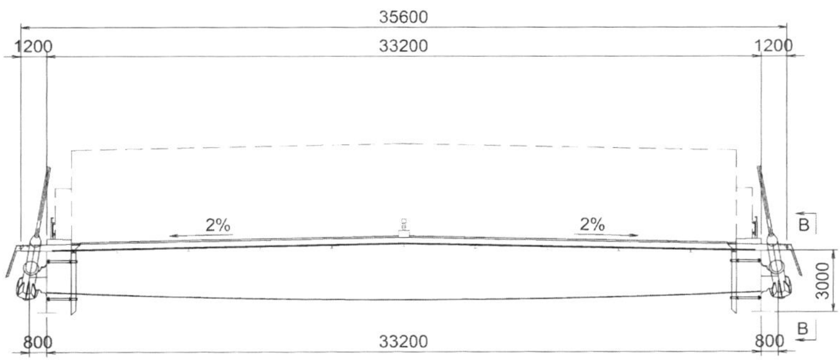

Fig. 4 Typical cross section of the girder

3620 pieces of pre-cast deck panel with 260 mm thickness are arranged on steel girder.

The bridge is large scale and there are few examples in the world for continuous cable stayed bridge.

2 Fabrication of Steel Structure

2.1 Fabrication Shop

Fabrication of steel girders and steel anchor boxes are distributed to 3 factories to reduce the fabrication time as in Table 1 [2].

For avoiding the dispersion of quality due to fabricating one continuous cable stayed bridge in 3 factories, IHI Aichi had initiative for other 2 factories. At first IHI Aichi fabricated the girder around Pylon 14 as the first erection part at construction site. After that, fabrication data made by IHI Aichi had been sheared with the other factories.

Table 1 Division of fabrication

Fabrication factory	Fabrication member	Fabrication total weight (ton)
IHI Corporation Co., Ltd. Aichi Works, Aichi, Japan (Hereinafter: "IHI Aichi")	Main girder, floor beam and support strut for P14	2500
IHI Infrastructure Asia Co., Ltd., Haiphong, Vietnam (Hereinafter: "IIA")	Main girder, floor beam and support strut for P13, P15 and P16	7600
Mitsui-Thang Long Steel Construction Co., Ltd., Hanoi Vietnam (Hereinafter: "MTSC")	Main girder, floor beam and support strut for P12 Anchor box for P12, 13, 14, 15 and P16	4400

Fig. 5 Fabrication of steel girder and trial assembling on IIA

Regarding stay cable anchorage of main bridge, it is very complex structure as shown in Fig. 5 and is required to weld the stiffener with over 40 mm thickness to web of I-beam. Especially because the welding between them is full penetration and this stiffener is installed diagonally, the appropriate welding sequence needs to be considered to minimize the deformation of web due to a lot of heat input welding.

3 General Erection Method

Three types of erection method (5 block erection, cantilever erection and closure erection) are applied as shown in the Fig. 6.

Erection unit of superstructure works is planned for each pylon, thus there are five work locations (P12, P13, P14, P15 and P16) in our site. Moreover, erection method of each pylon could be planned for similar procedure nevertheless there is the difference situation (water area or ground area) under the bridge because considering for transferring facility according to difference of erection area.

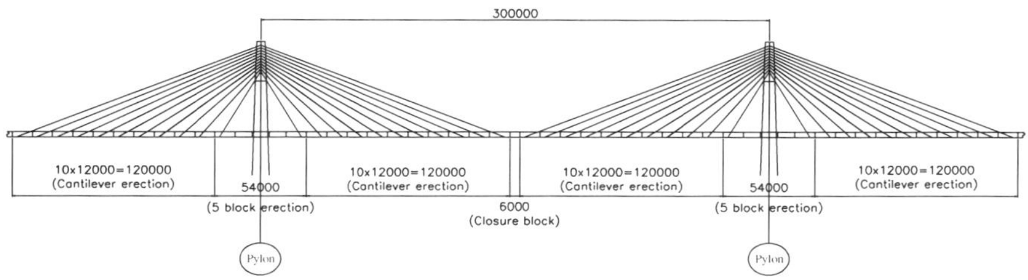

Fig. 6 Classification of erection method

4 5 Block Erection Around Pylon

Initial girder erection is performed at around pylon. 5 girder segments are erected by crawler crane in the condition that supported by temporary support system. After that, precast concrete deck slab is installed on the 5 girder segments and 1st Stay Cables support all weight of 5 block segment instead of Temporary support system. Controlling elevation of girder due to adjust the tension of 1st stay cable is performed. Finally casting joint concrete is performed.

Girder installation works is started after handing over the cross beam construction of pylon structure. Regarding girder installation work and construction work of upper pylon, they are carried out in parallel (Fig. 7).

4.1 Temporary Support System

Temporary support system is arranged for supporting dead load (W = 4000 kN) of steel girder and concrete deck slab around pylon. The support system is designed

Fig. 7 5 block erection around pylon and Installation of 1st Cable

as a triangle shape for transferring the vertical working force from dead load to horizontal force to the pylon structure.

After installing and tensioning of 1st stay cable, temporary support system is dismantled. C-shaped lifting balance frame which designed not to interfere with concrete deck slab and girder is used for dismantling.

5 Cycle Cantilever Erection

Balanced cantilever method is applied for erection of cable stayed bridge with adjusting the elevation of girder at each construction stage with prediction of bridge camber after completion and after 10 years later. To predict the bridge camber at several stages, it is important to make analysis model reflected construction stage and composite structure between concrete deck slab and steel girder. In this analysis model, even the fabrication error is considered (Figs. 8 and 9).

5.1 Working Platform

After 5 block erection around pylon, all the facilities are equipped to prepare for cycle cantilever erection such as 150 ton crawler crane and working platform. Working platform is travelling on Rail for maintenance vehicle with electrical trolley. Working Platform is designed as reducing the self-weight of working platform as much as possible to minimize the influence of girder. Crawler crane is travelling on Crane support rail that connected to floor beam with stud bolt directly (Fig. 10).

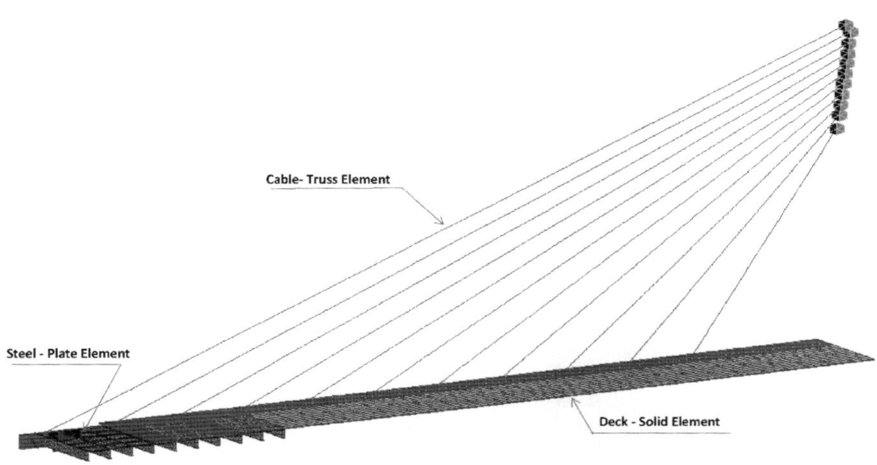

Fig. 8 FEM Model for predicting the crack occurrence

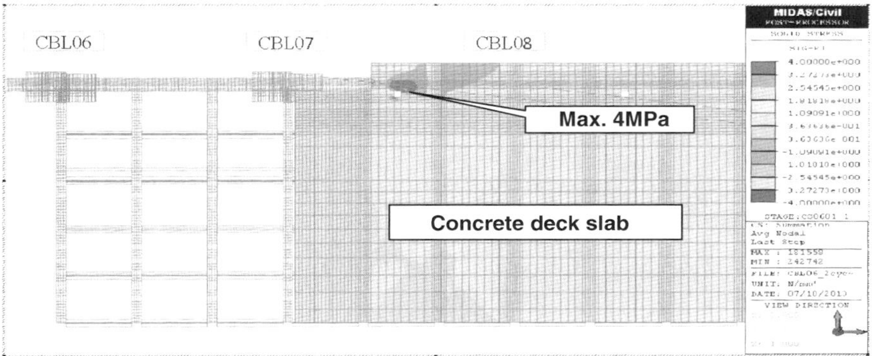

Fig. 9 Stress contour diagram

5.2 Construction Stage

10 times of cantilever erection are performed for each pylon including the following work items. It is planned at the period of 10 days per one cycle. Especially, to prevent the occurrence of crack on the concrete deck slab, the pre-cast deck panels are installed on the steel girder as much as possible after tensioning of the Stay Cable to reduce the tensile stress on the casted concrete deck slab.

- Steel girder installation and fastening high strength bolt
- Site painting on bolted joint
- Stay cable unreeling and installation
- Pre-cast deck panel installation
- Stay cable tensioning and adjusting girder elevation
- Joint concrete casting for concrete deck slab and curing
- Extension of crawler crane support rail and shifting working platform (Figs. 10 and 11)

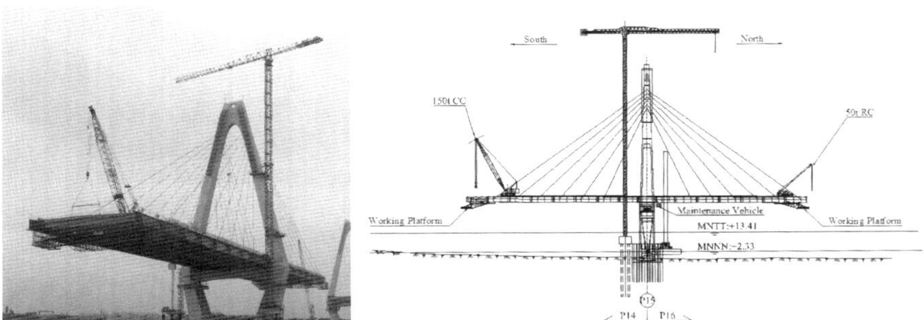

Fig. 10 Cycle cantilever erection

Fig. 11 Lifting up working platform and installation of working platform

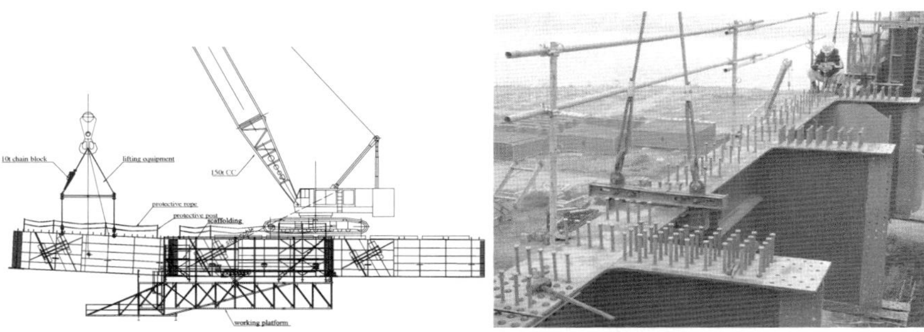

Fig. 12 Girder installation procedure and lifting clamp

5.3 Girder Installation

Edge girder and floor beam are installed by 150 ton crawler crane arranged on concrete deck slab. No lifting eye plate are arranged to reduce the total weight of steel. Thus lifting clamps are connected with upper flange of steel girder directly.

Center line of edge girder is controlled during installation with considering horizontal camber due to the self-weight of floor beam.

All splice joint are connected by Torque shear controlled high strength bolt (TCB M24 S10T) to shorten the time of connecting efficiently (Fig. 12).

5.4 Pre-cast Concrete Deck Panel and Joint Concreting

Pre-cast concrete deck panels are manufactured in north bank of project site. Manufacturing schedule is planned in accordance with bridge construction schedule because 6 months curing time is required after manufacturing (Fig. 13).

Fig. 13 Pre-cast concrete deck panel stock yard and installation on steel girder

Panel installation and casting joint concrete are carried out by 150 ton crawler crane on bridge deck. 100 % of strength of joint concrete is required before moving crawler crane for next cycle step in the Contract Technical Specification.

5.5 Stay Cable Installation and Tensioning

All stay cables are reeled horizontally in factory and unreeled by turn table from the end of girder erection to near the pylon on site.

Cable socket of pylon side is lifted up by tower crane and set into anchor box. Then on the socket of girder side, PC strand and tensioning rod are installed. Until the socket is set on the anchor position of girder side, it is pulled by 6000 kN center hole Jack (Fig. 14).

Tension force of stay cable is adjusted by thickness of shim plates. Final adjusting of girder elevation and tensile force of stay cable is carried out in night to stable the temperature of girder, deck slab, stay cable and pylon. For shortening the adjusting time, pre-tensioning is done in day time according to estimated calculation of tensile force. For checking of tensile force, self-vibration method is carried out with confirmation of pressure gauge of cable tensioning jack (Fig. 15).

Fig. 14 Stay cable unreeling and lifting up Stay Cable to install on pylon side

Fig. 15 Installation of stay cable (girder side) and adjusting shim plate

6 Bridge Closure

Bridge closure is performed at each span center after completion of cantilever erection of two pylons. Because main bridge is 6 spans continuous cable stayed bridge, there are 4 times of bridge closure and 2 times of end bridge closure in the Main Bridge. First and second bridge closure is not so difficult to close between cantilever segments because at least one side segment is free condition after completion of cantilever erection. However, after completion of end bridge closure, third and fourth bridge closure was very difficult and took time because of the condition fixed one side due to bridge closure of one time at last (Fig. 16).

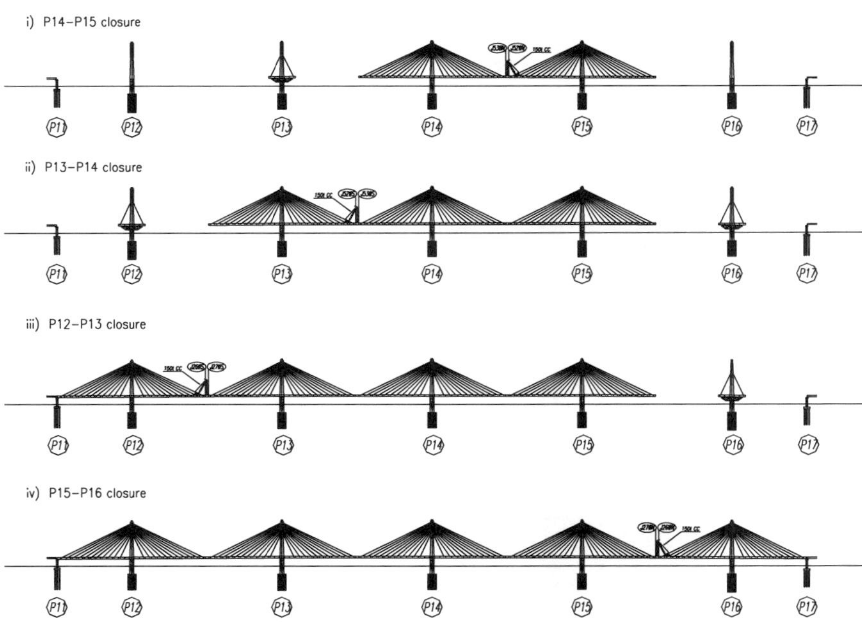

Fig. 16 Bridge closing procedure

6.1 Span Survey and Adjusting Closure Block

When remaining 3 segments of girder, span between tips of girders is surveyed then the length of closure girder block are decided and cut in accordance with survey result. To reduce the fabrication time of closure block, closure girder blocks are fabricated 200 mm longer than design value for each end in advance.

6.2 Alignment Adjusting

When closing the girder, to adjust the alignment between girders, Alignment adjusting device is applied. This alignment adjusting device is installed on two lower flanges of floor beam and pulled by two PC bars to close both girders. When difference of alignment was within 25 mm, alignment adjusting is performed by fixing splice plate of web plate and drifting of taper pin to bolt hole.

6.3 Elevation Adjusting

To adjust the elevation between two girders, concrete counter weight arranged on concrete deck slab are used before closing mainly. For the final adjustment, it performed by movement of 150 ton crawler crane or 50 ton Rafter crane.

6.4 Longitudinal Direction Adjusting

To adjust longitudinal direction between two girders, some 100 ton Jack arranged around bearing on the pylon is used. Because depending on the location of 150 ton crawler crane arranged on the concrete deck slab, jack load is difference, longitudinal direction adjusting is performed with adjusting girder elevation (Figs. 17 and 18).

Fig. 17 Installation of closure block and drifting pin to bolt holes

Fig. 18 Connection of closure block and connection of end closure block

7 Conclusion

On 15th April, 2014, Bridge Closing ceremony was held on bridge surface between P15 and P16 where final closure part of cable stayed bridge with many guests and all staffs relating for this project. The Nhat Tan Bridge will be opened to traffic on 4th January, 2015. The bridge is expected to further traffic access from Airport to Hanoi city and it will be helpful for the city's development.

Superstructure works spend 18 months for construction. During construction, we overcome challenges as below

- Due to control the elevation of deck for each construction stage in accordance with analysis result reflected construction stage, after closing the bridge, we did not have to adjust the cable tension force because all elevation value on the bridge is within allowable value.
- Due to implement of construction sequence decided by FEM analysis considered construction stage in previous, the crack over allowable width on the concrete deck slab can be prevented during 5 block erection, cantilever erection and closure works.
- Due to the strictly construction schedule and work control, we could reduce the construction period of superstructure for about 4 months in comparison with original schedule.

Acknowledgement We would like to thank for everyone who joined to the Nhat Tan Bridge Construction Project for their cooperation.

References

1. Kaifuku S, Kawashima H (2014) Closing of Nhat Tan Bridge (Vietnam-Japan friendship Bridge)-Construction of Package 1 for Nhat Tan bridge, Vietnam-. J Bridge Found Eng (Jpn) 48:46–47
2. Taki N, Kinoshita H, Yamashita M (2012) K, Matsuno, Construction of superstructure of Nhat Tan Bridge, Vietnam—Steel girder and concrete girder fabricated in Vietnam. J Civ Eng (Jpn) 53(12):29–32

Balanced Lift Method—A New Bridge Construction Technique

J. Kollegger and S. Foremniak

Abstract The economic performance of concrete bridges can be improved in certain design situations by using the balanced lift method, a new bridge construction method developed at the Vienna University of Technology. This method proposes to build the bridge girders in a vertical position and to rotate them into the final horizontal position with the aid of compression struts. The compression struts decrease the span of the bridge and allow lighter bridge girders, compared to bridges built by the balanced cantilever method or incremental launching, to be obtained. The balanced lift method can be applied for bridges with high and low piers enabling the construction of bridges over deep or low valleys. The first application of the new method is the construction of two bridges with low piers for the Austrian road management company (ASFINAG) crossing the rivers Lafnitz and Lahnbach on the S7 motorway in the south-east of Austria.

1 Balanced Lift Method

In conventional applications of precast concrete elements in bridge engineering, either match-cast elements are installed by using gantries or heavy bridge girders are lifted with mobile cranes [1, 2]. In a research project at the Vienna University of Technology a different approach was investigated called the balanced lift method. The main advantages of the balanced lift method are enabling an accelerated construction by using precast elements and decreasing the spans of a bridge through the usage of compression struts. The underlying idea was to construct the whole bridge in a vertical position and then rotate the bridge girders from the vertical into the final horizontal position with the aid of compression struts. A quick

J. Kollegger (✉) · S. Foremniak
Institute of Structural Engineering, Vienna University of Technology, Karlsplatz 13/E212-2, 1040 Vienna, Austria
e-mail: Johann.Kollegger@tuwien.ac.at

A. Caner et al. (eds.), *Developments in International Bridge Engineering*, Springer Tracts on Transportation and Traffic 9,
DOI 10.1007/978-3-319-19785-2_10

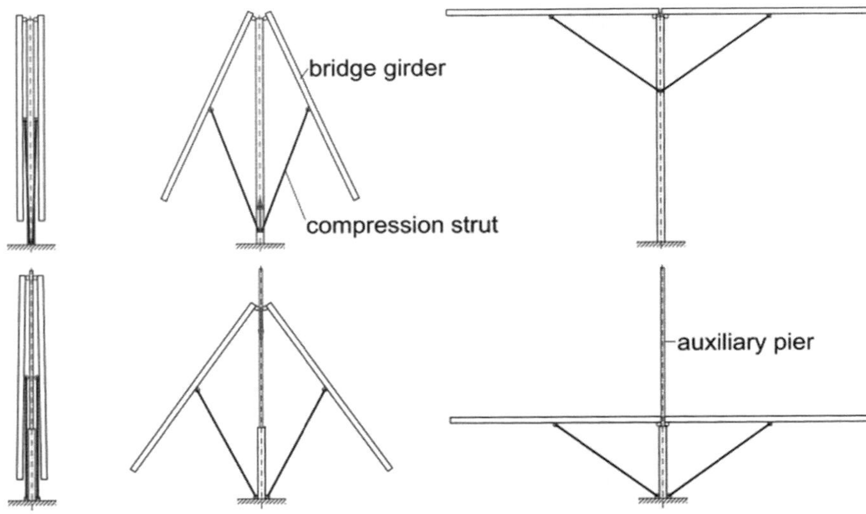

Fig. 1 Balanced lift method for high bridges *top*. Balanced lift method for low bridges *bottom*

way to describe this bridge building method is by comparing it to the process of opening an umbrella—with the bridge girders being the ribs and the compression struts being the stretchers. The compression struts are important load carrying members during the construction process and also in the finished bridge structure. As a result, much lighter bridge girders are obtained compared to bridges built by the balanced cantilever method or incremental launching.

The method has been primarily invented for bridges over deep valleys, meaning bridges with high piers, but can be—with minor changes—used for bridges with low piers. The example displayed in Fig. 1 (top) is suited for a valley bridge with high piers. The application of the balanced lift method for bridges with piers of small height is possible, if an auxiliary pier is used as is shown in Fig. 1 (bottom). The amount of force that is needed for the rotation process during the bridge construction depends, if one sets friction and eccentricities aside, on the weight of the bridge girders and the compression struts. Due to the importance of the weight of the different bridge elements the Institute for Structural Engineering at the Vienna University of Technology started working on different approaches for the usage of thin precast elements in collaboration with in situ concrete. The underlying idea is to minimize the weight of precast elements during transport and lift operations, then filling the elements with in situ concrete to create monolithic structures. Due to the very competitive costs compared to in situ concrete, the precise executions and the economic value, using precast elements reduces not only the construction period but also construction costs.

The international patent application procedures for the balanced lift method are running smoothly. The discussions with the examiners are restricted to formal issues and national peculiarities in the respective patent laws. Patents have been issued for Germany (DE 10 2006 039 551), USA (US 7,996,944), Russia (RU 2436890),

China (CN 1204627), Japan (JP5302195), Australia (AU2007288151) and Canada (CA2661311). The patent application procedures in Europe, Norway and India are still in progress. The first application of the new method is the construction of two bridges for the Austrian road management company (ASFINAG) crossing the rivers Lafnitz and Lahnbach on the S7 motorway in the south-east of Austria.

2 Application of Balanced Lift Method for Low Bridges

When designing bridges with low piers constructed with the balanced lift method one must keep in mind that the span of the balanced lift part must be held between 40 and 80 m [3]. Due to these small spans, a plate girder with a pre-fabricated bridge girder as shown in Fig. 2 is convenient as a cross section. In this case, the balanced lift method substitutes a construction either by incremental launching or by using falsework [4].

The design of bridges with piers of small height according to the balanced lift method will be described in this section for the example of the bridge across the river Lafnitz in the southeastern part of Austria. For the new S7 motorway "Fürstenfelder Schnellstraße" between Riegersdorf and the national border between Austria and Hungary, the rivers Lafnitz and Lahnbach must be crossed. The lengths of the Lafnitz Bridge and the Lahnbach Bridge are roughly 120 and 100 m, respectively. The cross section of the S7 motorway (Fig. 2) in this line section is traced out for two separate directed lanes, therefore the bridges across the rivers should be erected separately each with a width of 14.5 m, regarding prospective reconstruction measures.

The construction areas where the two bridges for the S7 motorway are to be built are ecologically sensitive and part of the nature reserve "Natura 2000". The bridges are basically needed to cross the rivers and to provide options for a deer pass. To avoid encroachment into the natural habitat, an erection on falsework is not accepted by the highway management company ASFINAG. The construction site should be as small as possible and kept to the central pier and abutments. To meet all these requirements a construction of the bridges would only be possible by the balanced cantilever method, incremental launching or the balanced lift method.

Before the alternative design using the balanced lift method was introduced, the plan was to build the bridges by incremental launching of steel bridge girders

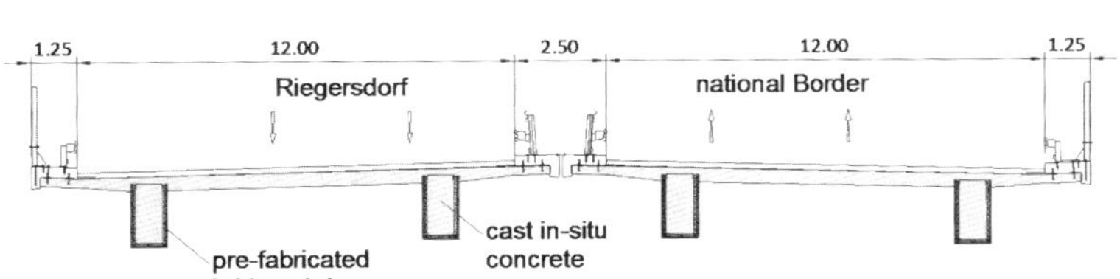

Fig. 2 Cross section of the S7 bridges

(Fig. 3). The cross section was, in order to withstand the bending moments during the launching process, very high compared to the cross section height achieved with the balanced lift method. The big difference in heights, 4.6 versus 2.0 m, can be achieved due to the compression struts which reduce the span lengths immensely (Fig. 4). The alternative design for the post-tensioned concrete bridges was based on a cross-section with a plate girder as shown in Fig. 2 [5]. It was proposed to build the central section of the webs by the balanced lift method as shown in Fig. 4, to install the end sections of the webs by mobile cranes placed behind the abutments, and to build the deck slab similarly to the original design by a formwork carriage. In the course of the preparation of the alternative design, the abutments and the locations of the central piers were rotated in plan by 30° with respect to the longitudinal axis of the bridge in order to react to the location of the riverbed and to provide an improved design for the deer pass. These changes resulted in a bridge design with two equal spans.

It could be shown that the construction costs for the post-tensioned concrete bridges erected with the balanced lift method amounted to only 70 % of the calculated costs of the composite bridges. When the Austrian highway management company ASFINAG became convinced of the financial benefits from a design based on the balanced lift method, a detailed design for the two bridges for crossing the rivers Lafnitz and Lahnbach was commissioned.

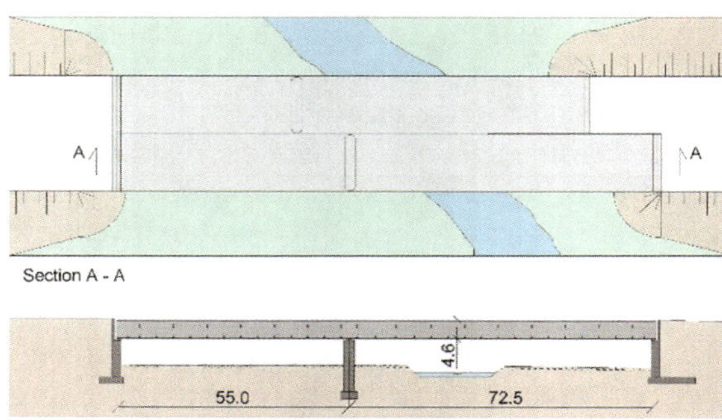

Fig. 3 Original design: a steel-concrete-composite bridge over the river Lafnitz

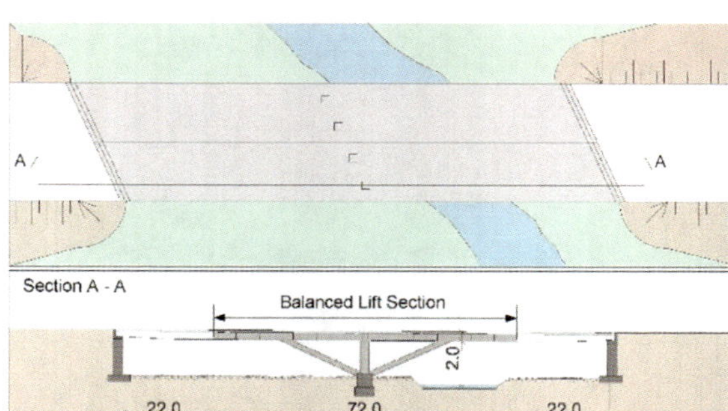

Fig. 4 Design based on the balanced lift method for a post-tensioned concrete bridge over the river Lafnitz

2.1 Large Scale Tests

During the design of the S7 bridges it was decided to construct a bridge test structure with a total length of 50.4 m. The design of the test structure was based on a 70 % scale of the design of the S7 bridges. Scaled down hollow reinforced concrete elements with small element thicknesses were prepared for the different elements of the bridge. For the 25 m long bridge girders, the side walls were 70 mm thick concrete elements. These elements were assembled on a steel form and then connected by a 120 mm thick slab of reinforced concrete. The height of the U-shaped sections of the 25 m long bridge girder was 1.26 m but the width varied from 700 mm to 1.4 m; the larger width was required at the point of connection with the compression strut.

The thin-walled bridge girder with its U-shaped cross section would have been too fragile for transport, assembly and the lifting process. Therefore, a truss made of reinforcing bars was welded to reinforcing elements protruding from the precast side wall elements at the top of the U-section. The U-section was hence converted to a box section which proved very robust in assembling and lifting operations. For the installation of tendons, the bridge girders were equipped with transverse concrete beams for future post-tensioning operations. Two 25 m long sections of a tower crane, equipped with guide rails, served as an auxiliary pier. Figure 5 shows the assembly of the compression struts in the vertical position. The bottom ends of the compression struts were equipped with 30 mm thick steel plates which were positioned quite accurately on concrete-filled steel tubes with an outer diameter of 150 mm, to provide an inexpensive hinge connection. The lifting of the second bridge girder is shown on the right of Fig. 5.

The lowering operations of the top points of the bridge girders with the aid of two mobile cranes are shown in Fig. 6. The maximum lifting force had been calculated at

Fig. 5 Auxiliary pier and assembly of compression struts with the aid of a mobile crane *left*. Assembly of the second bridge girder *right*

Fig. 6 Lowering of the top points of the bridge girders with the aid of mobile cranes

270 kN, which corresponded well with the lifting force measured by the cranes. On a real bridge, the compression struts would then be filled with cast in situ concrete. In this case, where the structure is intended as a demonstration project, only the node above the pier was filled with concrete to provide some wind resistance.

2.2 Designs According to the Balanced Lift Method for Bridges with Low Piers

In the last years, several additional conceptional bridge designs according to the balanced lift method have been prepared. As an example, the two bridges for the city bypass of Wieselburg are shown in Fig. 7. The cross-section for the bridges was designed as a plate girder with a width of 16 m, quite similar to the bridges on the S7 motorway described above. The location of the bridges is also situated in an environmentally sensitive area and it could also be shown to the owner that the application of the balanced lift method would have economic advantages compared to a design based on the incremental launching of steel girders.

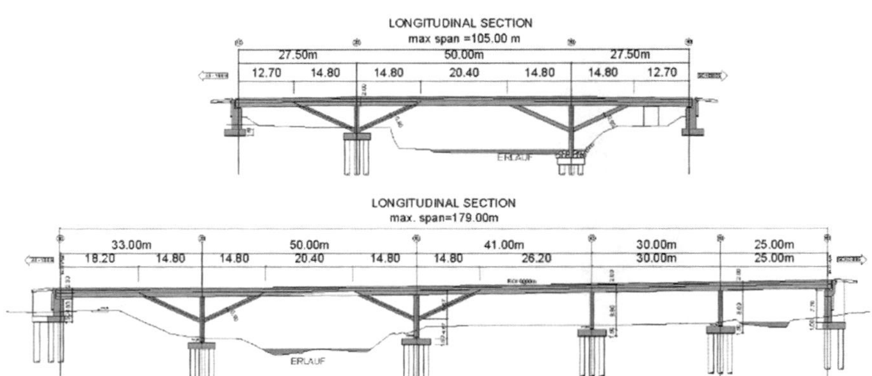

Fig. 7 Bridge design of two Erlauf bridges according to the balanced lift method

3 Application of Balanced Lift Method for High Bridges

When designing bridges with one high central pier constructed with the balanced lift method spans from 40 up to 125 m can be achieved. In this case the balanced lift method substitutes the balanced cantilever method. San Leonardo viaduct (Fig. 8) is a perfect example for a bridge that could have been constructed by the balanced lift method with one central pier only. The San Leonardo viaduct has two equal spans of 105 m. The bridge was built from 1968 to 1970 as part of the Palermo-Catania motorway in Sicily. If a deep and wide valley is spanned the balanced lift parts can be added enabling spans from 100 to 250 m (Fig. 9). In comparison to the plate girder cross sections of the bridges with low piers box girder cross sections are needed for spans larger than 100 m.

3.1 Bridge Girders Out of Hollow Wall Elements and Ultra-Thin Precast Elements

As it has been mentioned before the weight of the bridge girders and compression struts is of utmost importance during the lifting operation. Due to the large spans of the bridges bigger cross sections must be constructed. Therefore the cross sections as used for the S7 bridges could not be applied, but the idea of using precast elements in collaboration with in situ concrete was refined to create light box girders.

Fig. 8 San Leonardo viaduct in Sicily

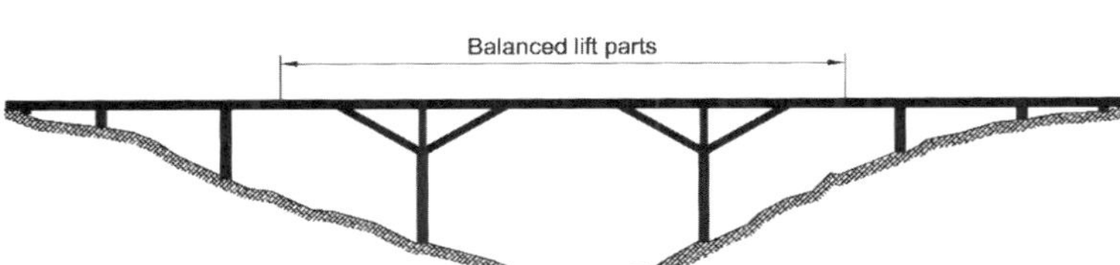

Fig. 9 A bridge across a deep, wide valley

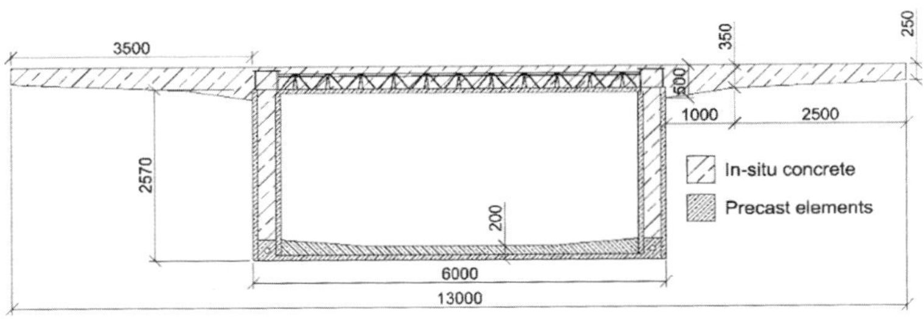

Fig. 10 Cross section of the bridge girder out of hollow wall elements and ultra-thin precast elements

Fig. 11 Reinforcement cages for the hollow wall elements. *Top* reinforcement cage using curved steel elements. *Bottom* reinforcement cage using lattice-girders

The webs of the bridge sections are made out of hollow wall elements and are held together by ultra-thin precast elements. In this state, the elements only have one fourth of the weight of a finished bridge girder with the same length (Fig. 10).

For a research project, two sections, each with a different kind of separation-reinforcement in the hollow wall elements, were built and tested. One set of hollow wall elements, which are used as the webs of the bridge girder, were manufactured using conventional lattice-girders (Fig. 11), the second set used curved steel elements (Fig. 11).

The construction of the precast elements starts with the manufacture of the webs. The hollow wall elements are manufactured fully automatically on a rotary production. Owing to the production process, the maximum width of the hollow wall elements is limited to 500 mm, the maximal height and length are limited to 3.40 and 12.00 m respectively. Due to limited space between the two slabs of the hollow wall element, all reinforcement and all of the ducts required for subsequent prestressing should be attached to the reinforcement cage before the hollow wall elements are cast.

While the wall elements are being manufactured the reinforcement cage for the floor slab and the deck slab can be produced. To reduce the weight of the precast section the floor slabs width is held to a minimum of 70 mm, with a stabilizing beam every two meters to strengthen the slab. The hollow wall elements are placed upright on a casting table and the floor slab including the stabilizing beam are cast

Fig. 12 Hollow wall elements are placed upright and the reinforcement cage for the floor slab is installed *left*. Casting of the floor slab *right*

Fig. 13 Deck slab on the casting table *left*. The finished precast bridge girder *right*

(Fig. 12), creating a u-shaped section. The length of the bridge girder segments is not limited by the length of the hollow wall segments (12 m). The hollow wall elements can be put side by side before the floor slab is cast, therefore creating a bridge girder segment with any desired length.

The deck slab is cast separately on a casting table with a stabilizing beam out of steel [Fig. 13 (left)]. After hardening the deck slab can be placed on top of the u-shaped section. Through welding parts of the deck slab and the stabilizing beam together with the u-shaped section, the precast bridge girder has reached its final state and is ready for transportation [Fig. 13 (right)]. Once the precast elements reach their destination and are installed in their final position, the filling of the floor slabs and the hollow wall elements with in situ concrete can begin. For the casting of the deck slab the thin precast element must be supported.

3.2 Large-Scale Tests on the Bridge Girders

Large-scale tests were used to provide valuable insight into the stability of the girders and for checking how precisely the weight of the actual girders corresponded with the calculated weights. The experimental girders were subjected to loading tests to assess their performance during transport and lift operations. The experimental setup of the large-scale tests is shown in Fig. 14.

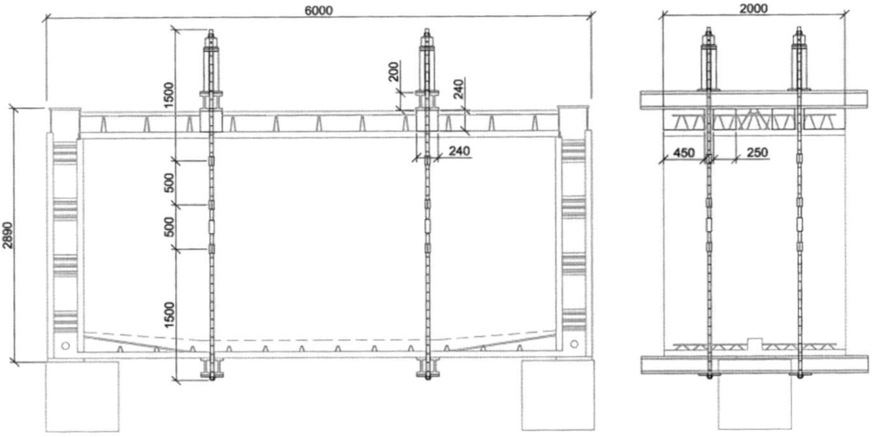

Fig. 14 Test setup in longitudinal and in profile

Fig. 15 Deflection of the deck slab during 5 load steps

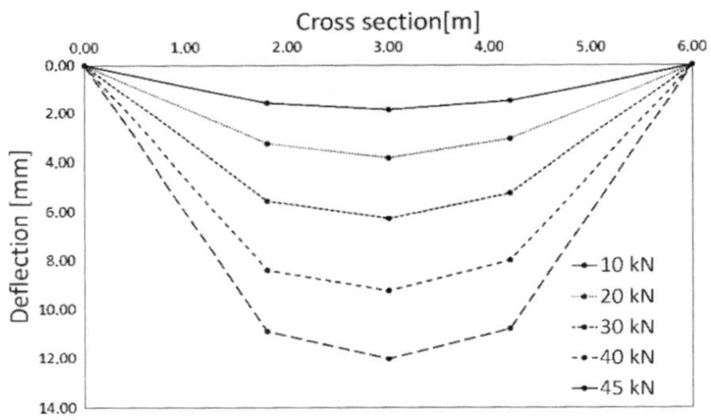

The bridge girder sections were placed on 3 bearings and the load was applied with four hydraulic jacks. In order to apply a linear load instead of a punctual load the jacks were placed on wooden beams. The deflection of the deck and floor slabs was measured with a set of linear variable differential transformers (LVDT's). Each precast bridge girder section was instrumented with 18 LVDT's, of which nine LVDT's measured the deflection of the deck slab and the other nine measured the deflection of the floor slab. Both precast bridge girder sections demonstrated the same load bearing behavior.

Since the only goal of the test was to provide an insight into the stability of the sections the experiment stopped at a load of 45 kN. At a load of approximately 45 kN the maximal measured deflection of the deck slab was approx. 12 mm. The mean of the deflections of the deck slab in correlation with the load is shown in Fig. 15. The stiffness of the precast bridge girder sections can be therefore calculated to be $EI = 7.9\ MNm^2$. The results confirm that the elements have enough stability to be transported and installed but will need extra support while the deck slab is being cast.

4 Conclusions

By employing the balanced lift method, the spans of the bridge girders are reduced by the compression struts, thus enabling considerable savings in construction materials. The proposed method will be especially advantageous for bridges with high piers and span lengths between 50 and 250 m. The usage of temporary piers enables an economic application of the balanced lift method for bridges with piers of modest height, as for example the two bridges on the S7 motorway.

Another advantage of the balanced lift method is the fact that all assembly and mounting operations are concentrated at the pier and that the rotation of the bridge girders can be carried out much faster than by horizontal launching of the bridge girders. The small space requirements and the high construction speed might be of advantage when an obstacle like a railway line or a busy motorway has to be spanned by a bridge and the interruption of the traffic routes has to be kept to a minimum. Different cross sections out of thin precast elements have been developed and tested allowing a construction of bridges with the balanced lift method in lengths between 50 and 250 m.

Acknowledgments The financial support of the large scale test by Österreichische Forschungsförderungs-gesellschaft (FFG), ASFINAG, ÖBB Infrastruktur AG and Vereinigung Österreichischer Beton- und Fertigteilwerke (VÖB) is gratefully acknowledged. The good cooperation with Schimetta Consult GmbH during the detailed design of the bridges for the S7 motorway, and with Franz Oberndorfer GmbH & Co KG during the construction of the large scale test structure in Gars am Kamp, is also gratefully acknowledged.

References

1. Rosignoli M (2013) Bridge construction equipment. ICE Publishing, London
2. Construction practices handbook for concrete segmental and cable-supported bridges (2008) American Segmental Bridge Institute
3. Gmainer S (2010) Brückenklappverfahren – Untersuchungen zur Entwicklung eines praxistauglichen Bauverfahrens. Dissertation, Vienna University of Technology
4. Wimmer D (2014) Entwicklung eines neuen Brücken- Bauverfahrens durch die Kombination von dünnwandigen Betonfertigteilen und Vorspannung. Dissertation, Vienna University of Technology
5. Wimmer D (2013) Thin-walled precast concrete girders for bridge construction and civil engineering. BFT International 79(8):44–51

Part III
Design for Extreme Events

Third Bosporus Bridge Aerodynamics: Sectional and Full-Aeroelastic Model Testing

Alberto Zasso, M. Belloli, T. Argentini, O. Flamand, G. Knapp, G. Grillaud, J.-F. Klein, M. Virlogeux and V. de Ville

Abstract The assessment of the aerodynamic performances of the Third Bosporus Bridge (BB3) has been realized through tests in two different Wind Tunnels, CSTB and Politecnico di Milano (POLIMI) and with different scale factors. The design process of a super long span bridge is strongly influenced by the wind actions on the bridge itself and the definitions of wind loads and wind induced dynamics cannot be done without relevant experimental campaigns. The definition of the local wind characteristics is the preliminary experimental test to be done, and the final verification of the bridge response to turbulent wind is the last stage of the wind design. For the Third Bosporus Bridge all these activities have been undertaken in close collaboration by the two cited laboratories. The collaboration granted a cross check of the results and of the test methodologies and hence assured a good reliability of the collected experimental data-base. In particular, at CSTB two sectional models with different scale factors 1:100 and 1:25 have been tested to assess the bridge stability, the response of the deck to vortex induced vibrations and to define the wind profile on the different lanes, and consequently the wind lateral loads on passing vehicles. Moreover, at CSTB a model of a tower has been tested in a very

A. Zasso (✉) · M. Belloli · T. Argentini
Politecnico di Milano, via La Masa 1, 20156 Milan, MI, Italy
e-mail: alberto.zasso@polimi.it

T. Argentini
e-mail: tommaso.argentini@polimi.it

O. Flamand · G. Knapp · G. Grillaud
CSTB—Centre Scientifique et Technique du Bâtiment, Nantes, France

J.-F. Klein
T-Engineering, Trollhättan, Sweden

M. Virlogeux
Consultant, Millau, France

V. de Ville
Greisch, Liège, Belgium

A. Caner et al. (eds.), *Developments in International Bridge Engineering*,
Springer Tracts on Transportation and Traffic 9,
DOI 10.1007/978-3-319-19785-2_11

large scale and high Reynolds number to understand the wind interaction of the stand-alone tower during the erection stages. At POLIMI the overall wind response have been tested using a full bridge 1:180 aeroelastic model to define the bridge response to turbulent wind and to check the bridge stability limit also during the erections stages. Finally a large 1:50 scale multi modal aeroelastic model simulating the torsional and vertical bending deck behavior was also tested in order to check in a larger scale possible vortex shedding induced vibrations as well as for a cross check with CSTB of possible Reynolds Number effects on the porous wind screens. The results of the experimental activities gave to the design team all the information needed to consider the wind response and highlighted the very good behavior of the bridge under wind actions from all the point of view: stability, vortex induced vibrations, wind loads and effects on the passing vehicles.

1 Introduction

The static and dynamic behavior of a bridge under wind actions is a fundamental aspect that allows one to assess the performance of the structure and the effectiveness of its design. The accurate study of bridge aerodynamics is a very important since many wind-structure interaction problems may occur: vortex-induced vibrations [1], aeroelastic instabilities (divergence, galloping, flutter). Numerical studies, performed in the preliminary design of the bridge, must be validated against wind tunnel tests on scale models before the final go for the project (e.g. [2–7]). In this paper, we present an overview of the wind tunnel testing procedure used to study the aerodynamics of the Third Bosporus Bridge. Wind tunnel tests were performed exploiting two complementary facilities, Politecnico di Milano (Polimi) and Centre Scientifique et Technique du Bâtiment (CSTB), in order to define a wide set of tests covering all the fundamental technical issues.

2 Step 1 of the Wind Tunnel Testing Procedure

The wind tunnel testing consisted of three subsequent steps. The first step is summarized in Fig. 1. It mainly consisted in the testing of the preliminary deck (static and dynamic) and in the selection of its optimal configuration (Fig. 2). Besides, towers were tested for static forces (Fig. 3). From pressure distribution analysis (Fig. 4), one can also see the vortex shedding phenomenon, which will be studied in step 2 using an aeroelastic model.

Due to the great width and slenderness of the bridge deck, carbon fiber was employed in the production of the aeroelastic sectional model. This has the advantage of a high stiffness and low weight and therefore provides a higher natural frequency in the deck model itself than more conventional materials, providing the best possible approximation to the assumption of an infinitely rigid sectional model.

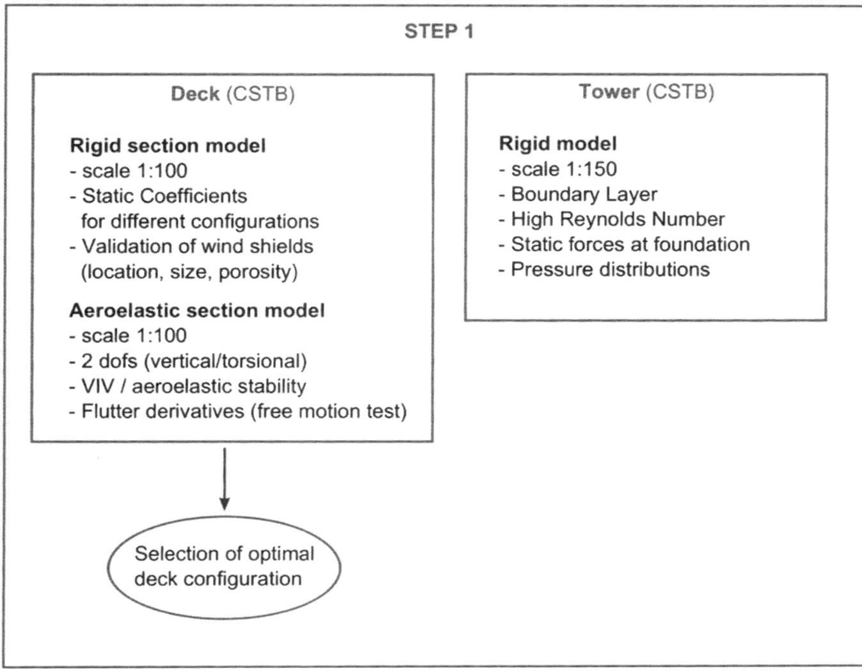

Fig. 1 Step 1 of the wind tunnel testing procedure

Fig. 2 CSTB 1:100 deck sectional model

The model was mounted upon a spring system and tested under low-turbulence conditions to examine the possibility of vortex-induced vibrations and to confirm deck stability under extreme wind speeds. The model was then retested in realistic turbulent wind conditions to quantify the static and dynamic behavior of the deck.

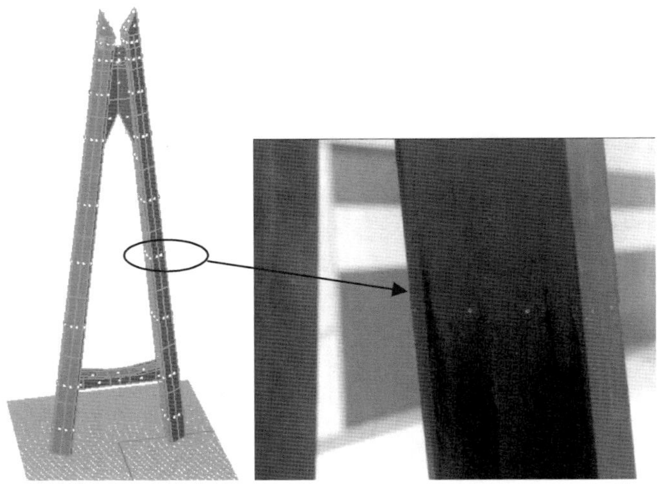

Fig. 3 Tower model tested in CSTB high Reynolds number wind tunnel. Pressure taps distribution and surfaces discretization (*left*)—Ring of pressure taps on one of the legs of the tower (*right*)

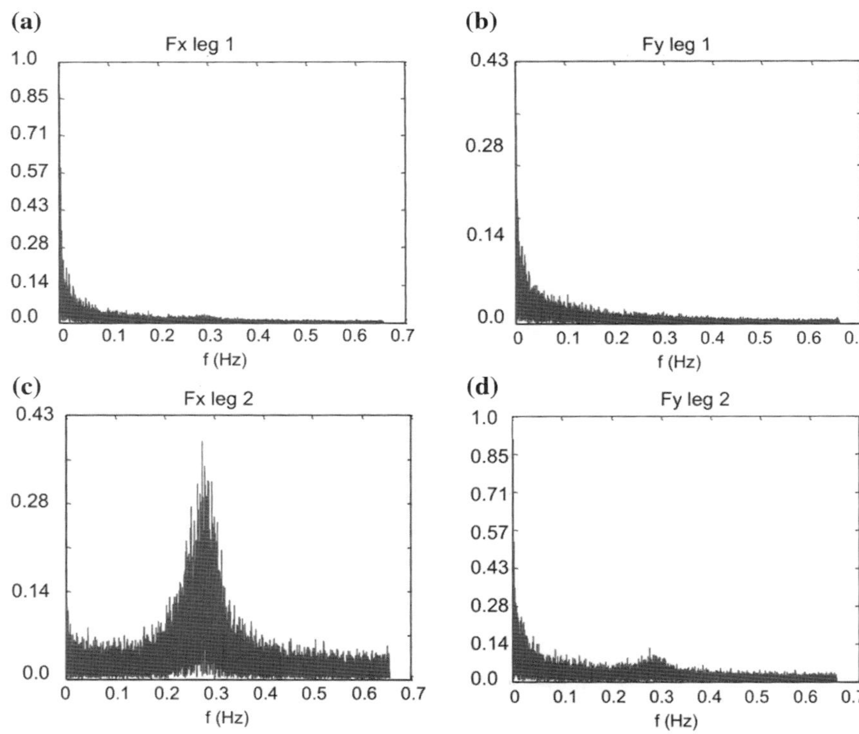

Fig. 4 Spectral analysis of the instantaneous X and Y normalized forces on leg 1 (**a** and **b** respectively) and leg 2 (**c** and **d** respectively) at 80° (full scale frequency) evidencing the vortex shedding phenomenon

The tower model was fabricated to a high degree of precision from carbon-fibres and epoxy resin. The tower was equipped with 229 pressure taps allowing measurement of local pressures at every point on the tower. The model was mounted on a high-frequency base balance to measure the overall dynamic load on the tower.

3 Step 2 of the Wind Tunnel Testing Procedure

The second step of the wind tunnel testing procedure was performed in the boundary layer wind tunnel of Politecnico di Milano. The tests performed are summarized in Fig. 5. A set of aeroelastic models in scale 1:180 were built to

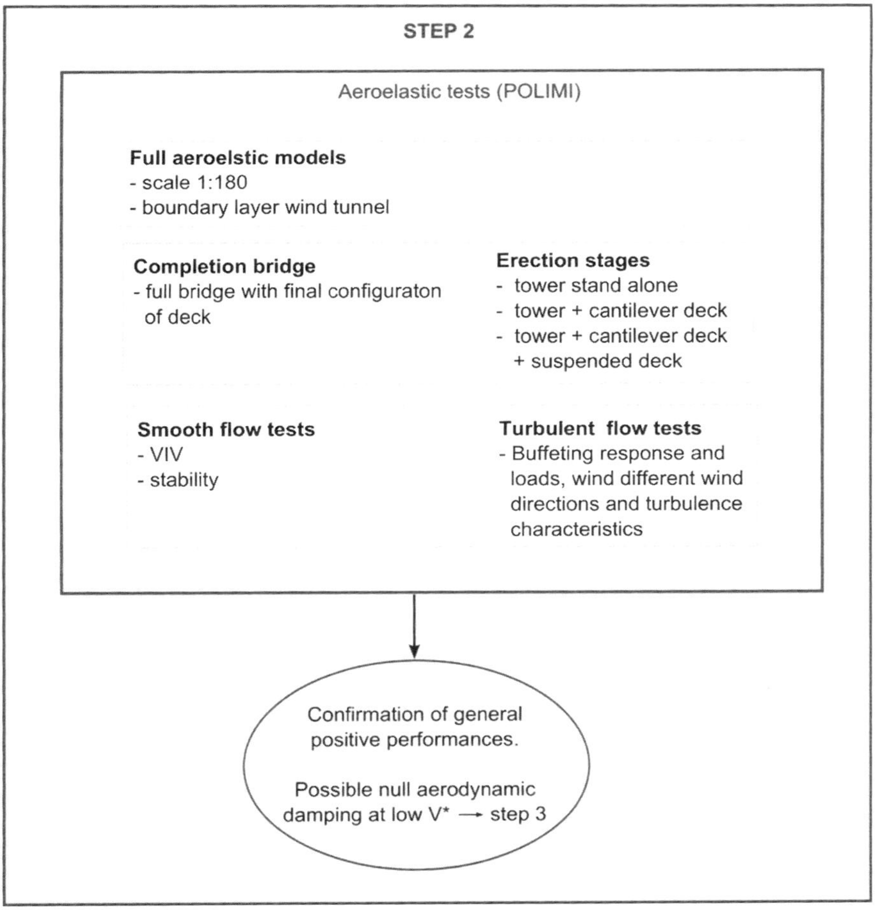

Fig. 5 Step 2 of the wind tunnel testing procedure

investigate vortex induced vibrations, aeroelastic stability, and buffeting response. Different wind directions, with different turbulence characteristics were tested.

Figures 6, 7 and 8 show 2 different aeroelastic models tested: full bridge, and the cantilevered-deck configurations. The models where designed using Froude similarity.

Typical results obtained from these kind of tests are reported in Figs. 9 and 10: the first plot reports the trend of the acceleration of two different sections of the deck in turbulent flow as a function of the wind velocity, measured with accelerometers. The second one shows the mean and standard deviation of the bending

Fig. 6 Aeroelastic model of the in-service configuration of the BB3 in the POLIMI wind tunnel

Fig. 7 Aeroelastic model of the cantilever deck construction stage of the BB3 in the POLIMI wind tunnel with turbulent flow conditions

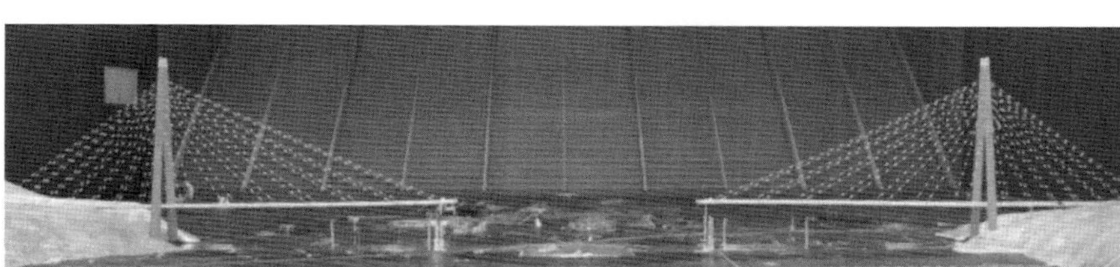

Fig. 8 Aeroelastic model of the cantilever deck construction stage of the BB3 in the POLIMI wind tunnel

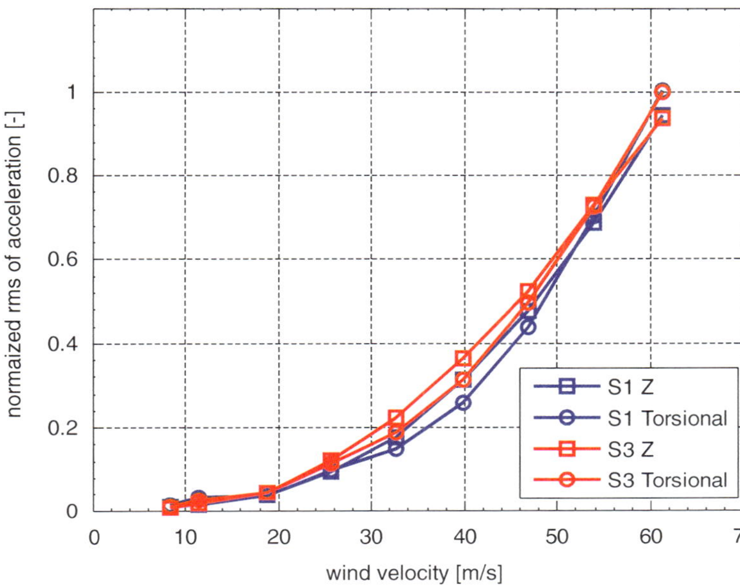

Fig. 9 Example of results from boundary layer wind tunnel tests: normalized vertical and torsional accelerations (reported at deck edge) as a function of wind velocity in turbulent flow

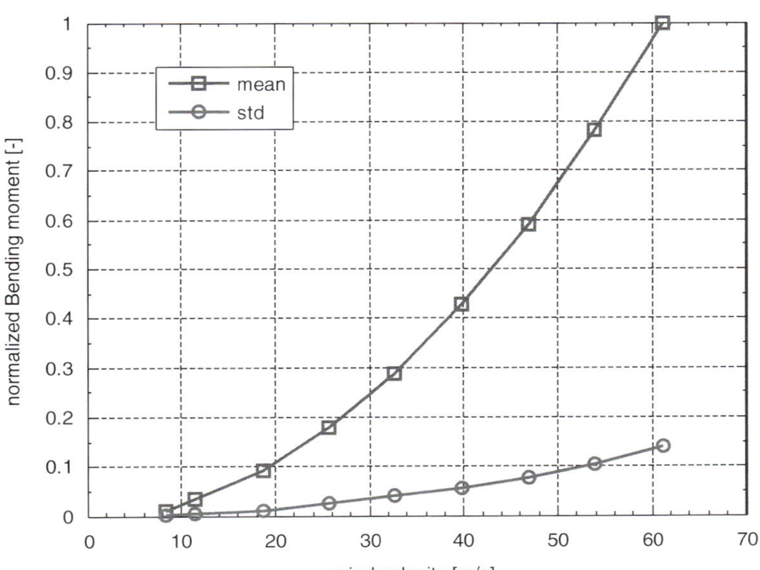

Fig. 10 Example of results from boundary layer wind tunnel tests: mean and std value of bending moment at tower foundations in turbulent flow as a function of the wind velocity

moment at the tower foundations in turbulent flow, measured with multi-component balances. The tests confirmed the general positive performances of the bridge at in-service and in construction stages.

4 Step 3 of the Wind Tunnel Testing Procedure

Since the deck aeroelastic coefficients have a torsional aerodynamic damping at low reduced frequency that is very low (nearly null), further investigations were made on larger scale models to check for VIV possible problems at higher Reynolds number and with more detailed models. Moreover, because it is made of half circular elements, the wind screen performance under Reynolds Number effects was tested. Two different models were tested as summarized in Fig. 11: at POLIMI an aeroelastic multi-modal deck model in 1:50 scale to study VIV (Figs. 12 and 13), at CSTB a 1:25 rigid deck section model to study the wind screens at high Reynolds Numbers (Fig. 14). Some comparisons between the models were done in terms of pressure distributions and vertical wind profiles to compare results with different Reynolds numbers. Figures 15 and 16 shows a comparison of the deck pressure distribution and of some vertical wind speed profiles between the two models that show a very good agreement.

These tests confirmed the good performances of the deck with respect to vortex induced vibrations, and allowed to have a fine characterization of the pressure field around the deck (to validate CFD models) and of the performances of different wind shields configurations.

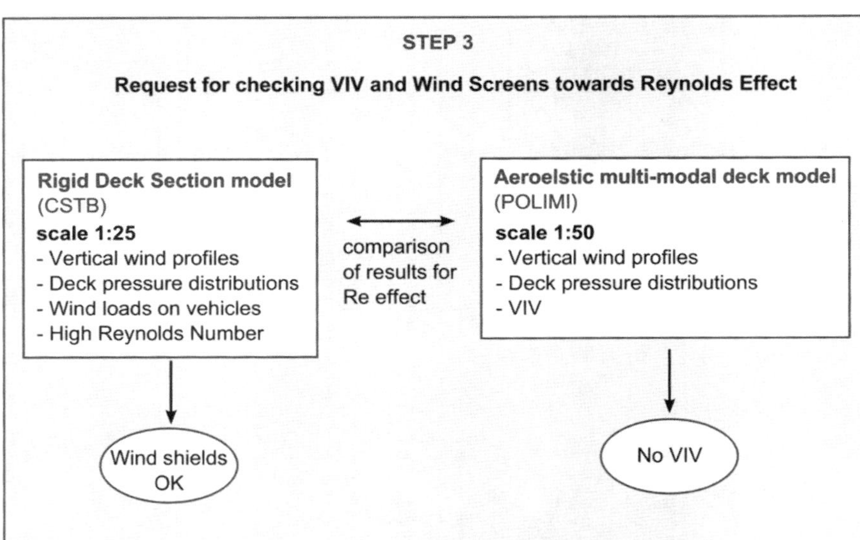

Fig. 11 Step 3 of the wind tunnel testing procedure

Fig. 12 Aeroelastic multi-modal deck model of the BB3 in the POLIMI wind tunnel scale 1:50

Fig. 13 Details of the aeroelastic multi-modal deck model of the BB3 in the POLIMI wind tunnel

Fig. 14 Rigid deck model of the BB3 in the CSTB wind tunnel scale 1:25

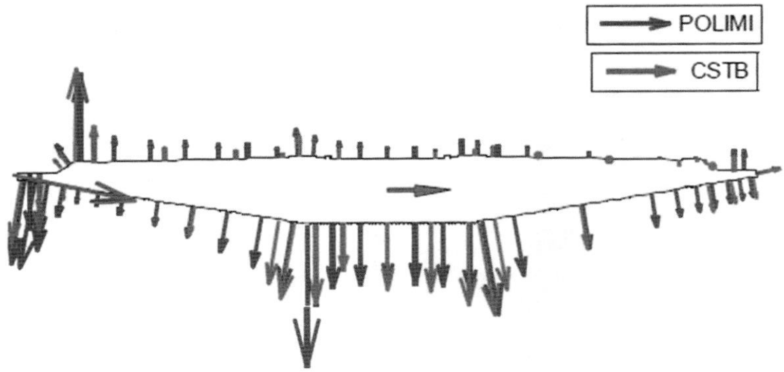

Fig. 15 Comparison of normalized mean pressure coefficients along the bridge deck: POLIMI and CSTB results

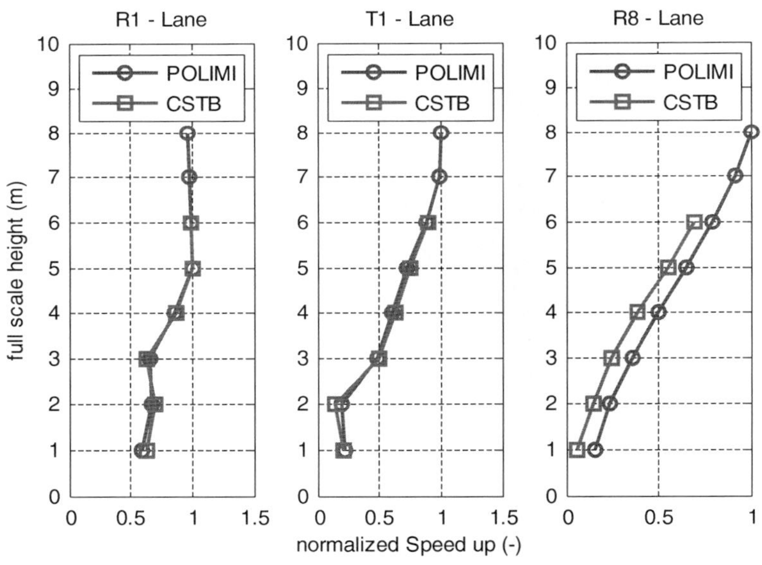

Fig. 16 Comparison of normalized wind speed up for a specific deck configuration with wind screens: POLIMI and CSTB results

5 Conclusions

The wind tunnel testing procedure for the Third Bosporus Bridge has been presented. The cooperation of two complementary wind tunnel laboratories (CSTB and POLIMI) has been exploited to perform a complete aerodynamic and aeroelastic characterization of the bridge.

The results of the experimental activities gave to the design team all the information needed to consider the wind response and highlighted the very good behavior of the bridge under wind actions from all the point of view: stability, vortex induced vibrations, wind loads and effects on the passing vehicles.

References

1. Belloli M, Fossati F, Giappino S, Muggiasca S (2013) Vortex induced vibrations of a bridge deck: dynamic response and surface pressure distribution. In: 12th Americas conference on wind engineering 2013, pp 256–265
2. Argentini T, Belloli M, Fossati F, Rocchi D, Villani M (2011) Experimental and numerical analysis of the dynamic response of cable-stayed bridge: vortex induced vibrations and buffeting effects. In: ICWE2011
3. Argentini T, Pagani A, Rocchi D, Zasso A (2014) Monte Carlo analysis of total damping and flutter speed of a long span bridge: effects of structural and aerodynamic uncertainties. J Wind Eng Ind Aerodyn 128:90–104
4. Diana G, Rocchi D, Argentini T (2013) An experimental validation of a band superposition model of the aerodynamic forces acting on multi-box deck sections. J Wind Eng Ind Aerodyn 113:40–58
5. Diana G, Rocchi D, Argentini T, Muggiasca S (2010) Aerodynamic instability of a bridge deck section model: Linear and nonlinear approach to force modelling. J Wind Eng Ind Aerodyn 98:363–374
6. Zasso A, Stoyanoff S, Diana G, Vullo E, Khazem D, Pagani KSA, Argentini T, Rosa L, Dallaire PO (2013) Validation analyses of integrated procedures for evaluation of stability, buffeting response and wind loads on the Messina Bridge. J Wind Eng Ind Aerodyn 122:50–59
7. Allori D, Bartoli G, Mannini C (2013) Wind tunnel tests on macro-porous structural elements: a scaling procedure. J Wind Eng Ind Aerodyn 123:291–299

Seismic Response of Bridges Considering Different Ground Motion Selection Methods

X. Liang, S. Günay and K.M. Mosalam

Abstract This paper makes use of different ground motion selection and scaling methods for the identification of predominantly first-mode engineering demand parameters (EDPs) of bridges under earthquake excitation. Two groups of ground motions are selected for this purpose. The first group, expected to result primarily in the first-mode response, is selected using the conditional mean spectrum (CMS) method. The second group, which serves as the reference for comparison, is selected to match a chosen scenario response spectrum. Since the shape of the chosen scenario spectrum allows higher mode response, the ground motions in this second group are considered as the ground motions with higher mode effects. Both groups of ground motions are selected using a method that seeks to match the mean and variance of the target spectrum. Comparison of the nonlinear time-history analysis responses from the two groups for three chosen ground motion scenarios indicates that higher mode effects are more pronounced on column displacements and deck accelerations than the column shear forces.

1 Introduction

The nonlinear structural response of bridge systems, similar to most complex structures, is intricate that it is often highly sensitive to the selection and modification of the input ground motions. This paper makes use of different ground motion selection and scaling methods for the identification of predominantly first-mode engineering demand parameters (EDPs) of bridges under earthquake excitation.

Two groups of ground motions are selected for this purpose. The first group, expected to result primarily in the first-mode response, is selected using the conditional mean spectrum, namely the CMS method. The second group, which serves

X. Liang · S. Günay · K.M. Mosalam (✉)
Department of Civil and Environmental Engineering, UCB, Berkeley, CA 94720-1710, USA
e-mail: mosalam@berkeley.edu

A. Caner et al. (eds.), *Developments in International Bridge Engineering*,
Springer Tracts on Transportation and Traffic 9,
DOI 10.1007/978-3-319-19785-2_12

as the reference for comparison, is selected to match a chosen scenario response spectrum. Since the shape of the chosen scenario spectrum allows higher mode response, the ground motions in this second group are considered as the ground motions with higher mode effects. Both groups of ground motions are selected using a method that seeks to match the mean and variance of the target spectrum. Nonlinear time-history analyses were conducted for selected highway bridges modeled in OpenSees [1] using the two groups of ground motions. Comparison of the nonlinear time-history analysis responses from the two groups for three chosen ground motion scenarios indicates that higher mode effects are more pronounced on column displacements and deck accelerations than the column shear forces. Therefore, the column base shear is likely a first-mode-dominant EDP.

2 Bridge Structures and Analytical Models

Three highway bridge structures, reflecting common practice in California, are used in this study. These bridges are: (1) *Jack Tone Road Overcrossing* (Bridge A), a bridge with two spans supported on a single-column bent, (2) *La Veta Avenue Overcrossing* (Bridge B), a bridge with two spans and a two-column bent, and (3) *Jack Tone Road Overhead* (Bridge C), a bridge with three spans and two three-column bridge bents. Extensive simulations were conducted on these bridges [2] using OpenSees [1], on which the modeling assumptions adopted herein are based. OpenSees [1] has a sufficient element and material response library and empowers scripted execution of repetitive nonlinear time-history analyses (NLTA) through which the model parameters and input ground motions can be systematically varied. The bridge models used in the simulations are depicted in Figs. 1 and 2. Seat-type abutments, shear keys, expansion joints, column-bents, and the superstructure are included in the model. For a detailed explanation of the employed modeling assumptions, refer to the descriptions in [3].

Two approaches are considered for abutment modeling: detailed and simplified. In both approaches, the longitudinal responses of the backfill and the expansion joint, the transverse responses of the shear keys, and the vertical responses of the bearing pads and the stemwall are all explicitly considered. In the detailed modeling approach (Fig. 3a), five nonlinear springs, connected in series to gap elements, are used to model the passive backfill response and the expansion joint. The strength and initial stiffness of the soil springs are determined according to Caltrans SDC [4]. The shear key response is modeled using a nonlinear spring with a tri-linear backbone curve. The vertical response of the bearing pads and the stemwall is modeled by two parallel springs that represent the stiffness of the bearing pads and the stemwall. In the simplified modeling approach, the number of nonlinear springs connected in series to the gap elements is reduced to two as shown in Fig. 3b, and the shear key response is modeled using an elastic-perfectly-plastic backbone curve.

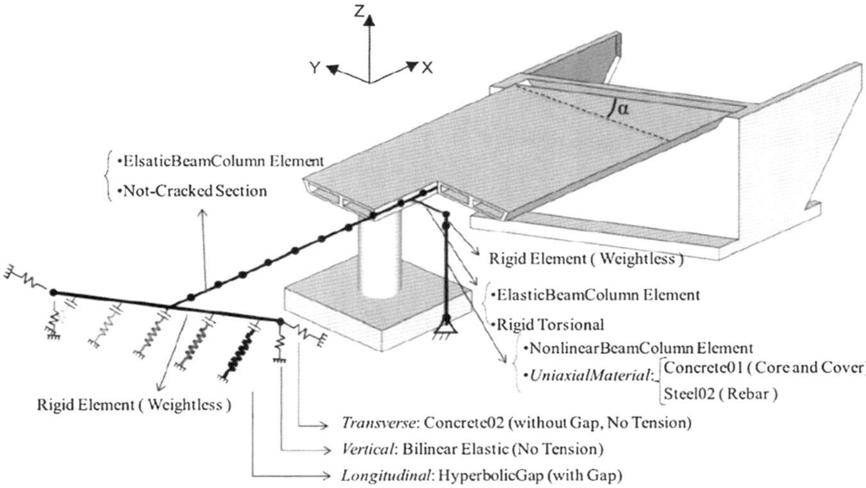

Fig. 1 Modeling of bridge B [2]

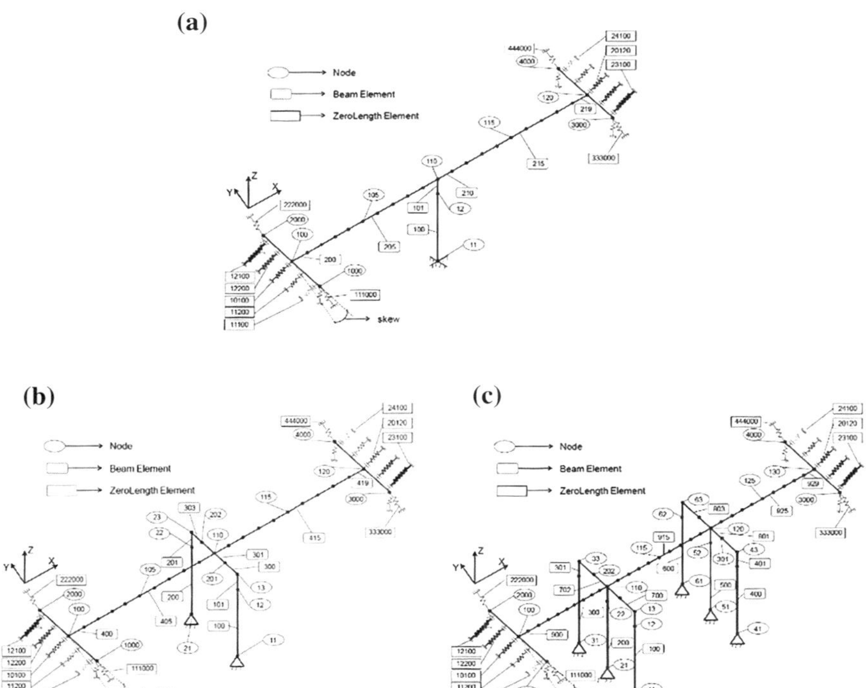

Fig. 2 Modeling of the analyzed bridges [3]. **a** Bridge A, **b** bridge B, **c** bridge C

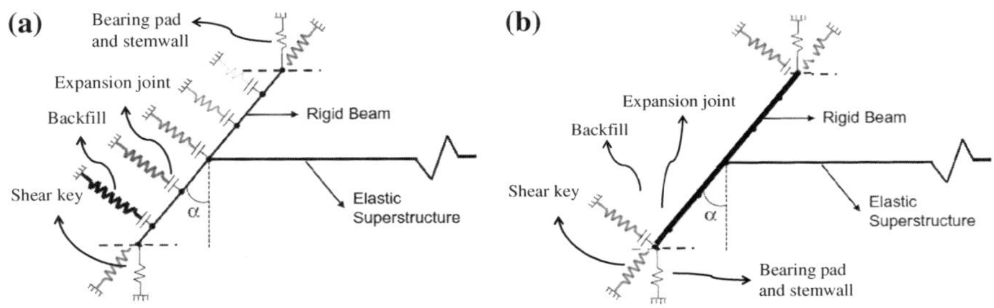

Fig. 3 Springs and gap elements used to model various components of the abutment [3]. **a** Detailed modeling, **b** simplified modeling

3 Ground Motion Selection

For each bridge, three earthquake scenarios are considered, namely those with 2, 10 and 50 % probabilities of exceedance (POE) in 50 years. The attenuation model by Campbell and Bozorgnia [5] is used to approximate these three hazard levels. Figure 4 shows Uniform Hazard Spectra (UHS) of the three hazard levels and their approximations for the site of Bridge B.

The CMS [6], which is a response spectrum associated with a target value of the spectral acceleration *Sa* at a single period, serves as the target spectrum for the first group of ground motions. In this study, this single period is the fundamental period of the considered bridge. The second group, which serves as the reference for comparison, is selected to match the spectrum predicted by the attenuation model of Campbell and Bozorgnia [5]. Figure 5a shows the response spectrum by the attenuation model from Campbell and Bozorgnia [5] that approximates the UHS at hazard level of 10 % POE in 50 years for Bridge B, i.e. the target spectrum for the second (reference) group. Also shown in Fig. 5a is the CMS [6] anchored at the fundamental period of Bridge B of 1.1 s, which is the target spectrum for the first group. As mentioned before, both groups of ground motions are selected using a method proposed by Jayaram et al. [7] that seeks to match the mean and variance of the target spectrum (Fig. 5b). With a target distribution, Monte Carlo simulation is used to probabilistically generate multiple realizations of response spectra, and then recorded ground motions whose response spectra individually match the simulated response spectra are selected. Furthermore, a greedy optimization is applied to improve the match between the target and the sample means and variances. This is performed by replacing one previously selected ground motion at a time with a record from the database that generates the best improvement in the match. For a detailed explanation of this method, refer to [7].

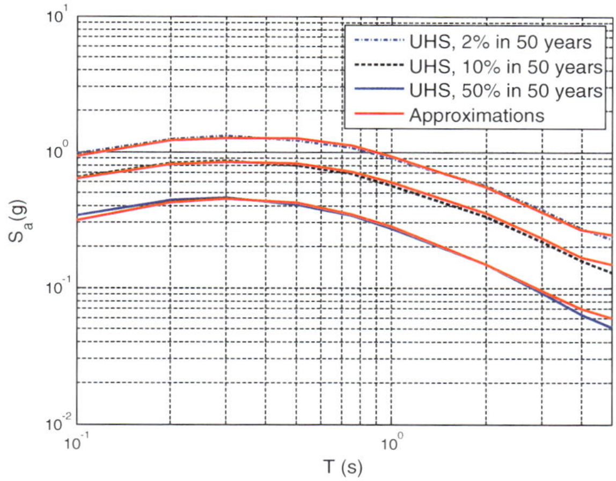

Fig. 4 Response spectra of three hazard levels for bridge B site

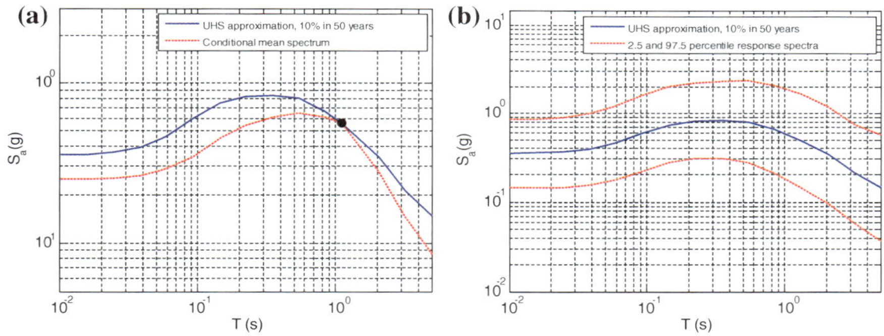

Fig. 5 Uniform hazard spectrum and CMS for 10 % POE in 50 years for the site of Bridge B. **a** Median, **b** median and variance

4 Simulation Results

As discussed above, there are two approaches for abutment modeling: detailed (Fig. 3a) and simplified (Fig. 3b). For each of the two abutment modeling of each of the three bridges, 40 ground motion records are selected for each of the three scenarios per each of the two ground motion groups, and NLTA are conducted with the selected ground motions, which yield a total of $2 \times 40 \times 3 \times 2 = 480$ simulations per bridge for the three considered bridge systems. The maximum column drift ratio, base shear, and deck total acceleration are selected as the EDPs.

Figures 6, 7 and 8 present the ratios of the median EDPs from the ground motions of the first group (CMS) to those from the ground motions of the second reference group. Because the first and second groups are the first-mode dominant

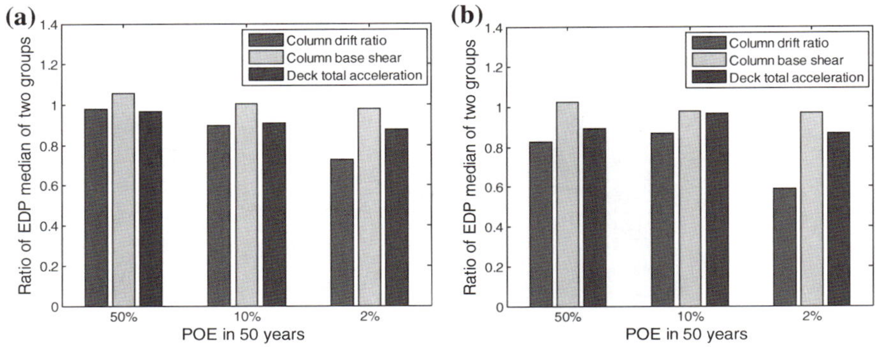

Fig. 6 Ratios of median EDPs for the two abutment modeling cases of bridge A. **a** Simplified modeling, **b** detailed modeling

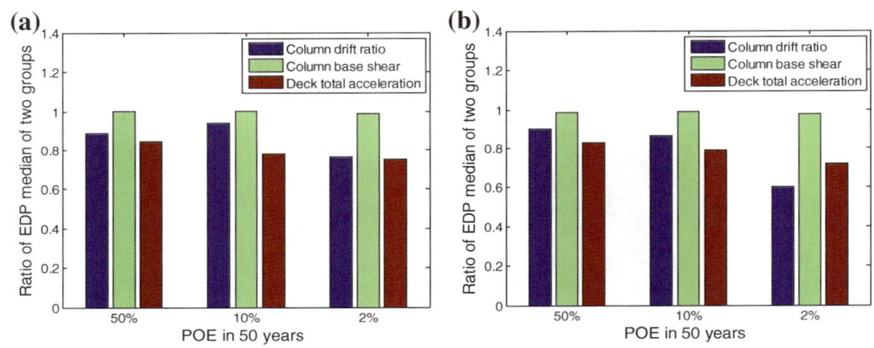

Fig. 7 Ratios of median EDPs for the two abutment modeling cases of bridge B. **a** Simplified modeling, **b** detailed modeling

and higher mode response reflecting ground motions, respectively, the smaller the ratio, the more the considered EDP is affected by the higher modes. It is observed that the ratio for the column base shear is close to 1.0 and almost invariant for both modeling cases and the three scenarios. On the other hand, the ratios for the column drift and deck acceleration are always less than 1.0 and reduce as the hazard level and the corresponding nonlinearity level increase. Accordingly, it is concluded that the higher mode effects are more pronounced on column displacements and deck accelerations than on column shear forces. Moreover, the effects of higher modes on the column drift and deck acceleration increase with increasing nonlinearity. However, the column base shear is likely to be a first-mode dominant EDP, independent of the hazard level.

For the 10 and 2 % POE in 50 years cases, the first mode dominancy of the column base shear can be explained by the fact that it is limited by the base shear capacity of the column (direct shear or the shear force limited by flexure).

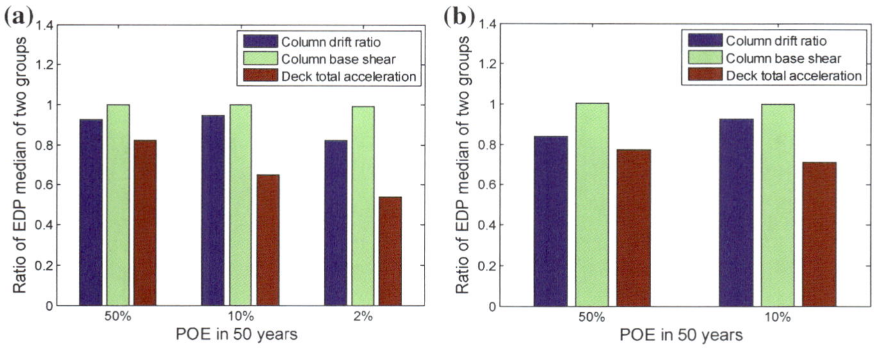

Fig. 8 Ratios of median EDPs for the two abutment modeling cases of bridge C. **a** Simplified modeling, **b** detailed modeling

However this explanation is not valid for the 50 % POE in 50 years scenario, where the response is below the column base shear capacity, indicating that the column base shear is not affected from higher modes also in the linear range.

It is noted that the comparison is not recorded for Bridge C with detailed abutment modeling for the hazard level of 2 % POE in 50 years. This is due to the fact that 7 out of 40 ground motion records in the second group failed to converge, despite significant efforts to improve convergence [8].

The results in Figs. 6, 7 and 8 are supportive of the statement that the higher modes affect the response of bridges to a greater extent than that of buildings as indicated in [9]. Furthermore, examining the obtained results from a ground motion selection aspect, it can be stated that the CMS method underestimates the column drift ratio and the deck total acceleration with respect to the ground motions selected to match the UHS. In order to evaluate the accuracy of these methods, an estimate of the true response can be obtained using the concept of high-end prediction (HEP) [10, 11]. However, this investigation is beyond the scope of the presented study.

5 Summary and Conclusions

This study employed different ground motion selection and scaling methods to identify the predominantly first-mode engineering demand parameters (EDPs) of bridges under earthquake excitation. For this purpose, the median maximum responses of three selected EDPs, namely column drift ratio, the column base shear, and the deck total acceleration, from two groups of ground motions, were investigated. The responses from the first group, selected using the conditional mean spectrum, were expected to represent the first-mode-dominant response. The responses from the second group, with the spectrum shape allowing higher mode responses, were considered as responses including the higher mode effects.

A method that seeks to match both the mean and variance of the target spectrum was utilized for selection of ground motions for the two groups. Comparison of the responses from 1440 nonlinear time-history analyses obtained from the two groups for three representative ground motion scenarios, i.e. 50, 10 and 2 % probability of exceedance in 50 years, on three bridge systems with two approaches to model their abutments, indicated the higher mode effects are more pronounced on column displacements and deck accelerations than on column shear forces, the effect of higher modes increases with increasing nonlinearity and the column base shear is a first-mode-dominant EDP, independent of the hazard level.

Acknowledgments This research was supported by Caltrans under Contract Number 65A0454 for the Project "Guidelines for Nonlinear Seismic Analysis of Ordinary Bridges: Version 2.0". The authors thank Prof. F. Zareian, UC Irvine, for providing the OpenSees models.

References

1. McKenna F, Fenves GL, Filippou FC, OpenSees, 2010
2. Kaviani P, Zareian F, Taciroglu E (2012) Seismic behavior of reinforced concrete bridges with skew-angled seat-type abutments. Eng Struct 45:137–150
3. Kaviani P (2011) Performance-based seismic assessment of skewed bridges. Ph.D. dissertation, University of California, Irvine
4. Caltrans SDC (2010) Caltrans seismic design criteria, version 1.6, California Department of Transportation, Sacramento
5. Campbell KW, Bozorgnia Y (2008) NGA ground motion model for the geometric mean horizontal component of PGA, PGV, PGD and 5 % damped linear elastic response spectra for periods ranging from 0.01 to 10 s. Earthq Spectra 24:139–171
6. Baker JW (2011) Conditional mean spectrum: tool for ground motion selection. ASCE J Struct Eng 137(3):322–331
7. Jayaram N, Lin T, Baker JW (2011) A computationally efficient ground-motion selection algorithm for matching a target response spectrum mean and variance. Earthq Spectra 27(3):797–815
8. Liang X, Günay S, Mosalam KM (2014) Integrators for nonlinear response history analyses: revisited. In: Proceedings of the Istanbul bridge conference, Istanbul, Turkey
9. Kappos AJ, Gkatzogias KI (2013) Gidaris IG extension of direct displacement-based displacement-based design methodology for bridges to account for higher mode effects. Earthq Eng Struct Dynam 42:581–602
10. Haselton CB, Baker JW, Bozorgnia Y, Goulet CA, Kalkan E, Luco N, Shantz TJ, Shome N, Stewart JP, Tothong P, Watson-Lamprey JA and Zareian F (2009) Evaluation of ground motion selection and modification methods: predicting median interstory drift response of buildings technical report 2009/01, Pacific Earthquake Engineering Research Center, University of California, Berkeley
11. Mobasher, B, Junca PP, Zareian F, Haselton CB (2013) Sensitivity of response of seat-type abutment bridges to ground motion selection and scaling methods. In: 10th international conference on urban earthquake engineering, Tokyo

A Proposal for the Improvement of the Earthquake Resistance of Multi-span Precast I-Beam Bridges

Ioannis A. Tegos, Sevasti D. Tegou and Mihail A. Tsitotas

Abstract This study proposes an innovative restraining system for the improvement of the earthquake resistance of precast concrete I-beam bridges mainly in the longitudinal direction. The proposed method is based on the connection of the continuous deck slab of the bridge through the sidewalks with both abutments and the elimination of the end expansion joints. In this way, the bridge obtains four restraints (one per wing-wall), which behave as tension ties during the deck contraction and have the ability to reduce the seismic movements of the deck. The resultant in-service requirements due to creep, shrinkage and thermal effects are properly accommodated. Each sidewalk is divided to zones that are anchored to the deck and zones that have the ability to slip on the deck. The efficiency of the proposed method is investigated by utilizing a precast I-beam bridge of Egnatia Odos Motorway in Greece. This bridge has been converted to a ductile bridge system by using active seismic stoppers at the head of the piers.

1 Introduction

Conventional bridges are supported on piers and abutments through bearings, while expansion joints (between the deck of the bridge and the abutment) are provided in order that length changes of the bridge's deck—due to temperature and other effects—take place without inducing additional stress and strain fields. This way, the system abutment-embankment is not a part of bridge's structural system and therefore it does not play an active role in resisting the earthquake action.

A critical part of the aforementioned bridge systems is the sensitivity to seismic displacements [1]. In high seismicity regions, precast I-beam bridges with tall piers

I.A. Tegos · S.D. Tegou (✉) · M.A. Tsitotas
Department of Civil Engineering,
Aristotle University of Thessaloniki, Thessaloniki, Greece
e-mail: stegou@civil.auth.gr

A. Caner et al. (eds.), *Developments in International Bridge Engineering*,
Springer Tracts on Transportation and Traffic 9,
DOI 10.1007/978-3-319-19785-2_13

that are connected to the deck through elastomeric bearings usually develop large displacements. According to Code's provisions, they could be limited by utilizing energy dissipation devices [2–5].

However, it is possible to include stoppers between the piers and the deck of the bridge, Fig. 1, in order to enforce common displacements in these elements. Actually, modern earthquake codes, such as Eurocode 8- Part 2 [6] include a methodology for designing bridges with stoppers acting as restrainers. Nevertheless, this methodology is followed very rarely, since designers prefer to include adequate gaps to address the creep, shrinkage and thermal effects.

On the other hand alternative methods that lead to a reliable and cost effective design have been proposed in recent studies [7–10]. Recently, there have been many efforts aiming at reducing the seismic displacements of bridges. In particular, Mikami et al. [11] studied ways of including the abutments in the bridge's resistance to earthquakes. Nutt and Mayes [12] showed that this way the cost of the bridge is reduced. Along these lines there has been recently a lot of activity at the Reinforced Concrete and Masonry Structures Lab of the Aristotle University

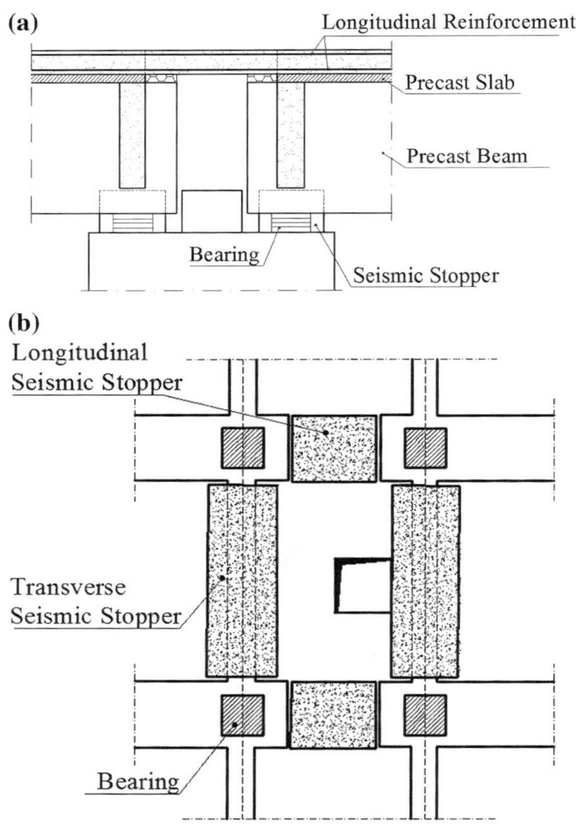

Fig. 1 **a** Active longitudinal seismic stopper at pier of a precast I-beam bridge and **b** plan view of a pier's head

of Thessaloniki, Greece. Many studies had been made on the issue of the use of embankments, abutments for the enhancement of the earthquake resistance capability of the structure [7, 8].

This work is an attempt at studying the effects of including the sidewalks in the structural system of the bridge against earthquake action. In particular, in this study is investigated how the sidewalks may work in as restrainers. The proposed methodology is reliable and cost-effective, and it is proposed as an alternative to the base isolation one that has started been widely accepted in recent years.

2 Description of the Proposed Restraining System

Bridge sidewalks are non-structural elements that are included in the additional permanent loads of the bridge. In this study, the sidewalks are developed and are considered to participate in the earthquake resisting system of the bridge.

The end parts of the bridge sidewalks are converted to seismic restrainers, which behave as tension ties and restrain the longitudinal movements of the bridge. The so-called seismic active length of the sidewalks is mainly determined by the serviceability requirements of the bridge. The configuration of the proposed restraining system is given in Figs. 2 and 3. Each sidewalk is divided to three zones. The outer zones are anchored to the deck while the central one has the ability to slip on the bridge deck, Fig. 3b.

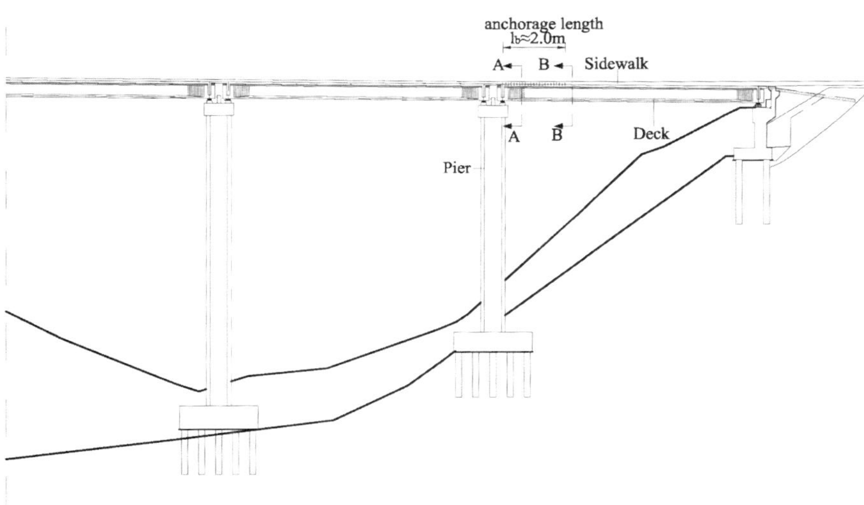

Fig. 2 Longitudinal section of a precast I-beam bridge in which the proposed sidewalks-restrainers are implemented

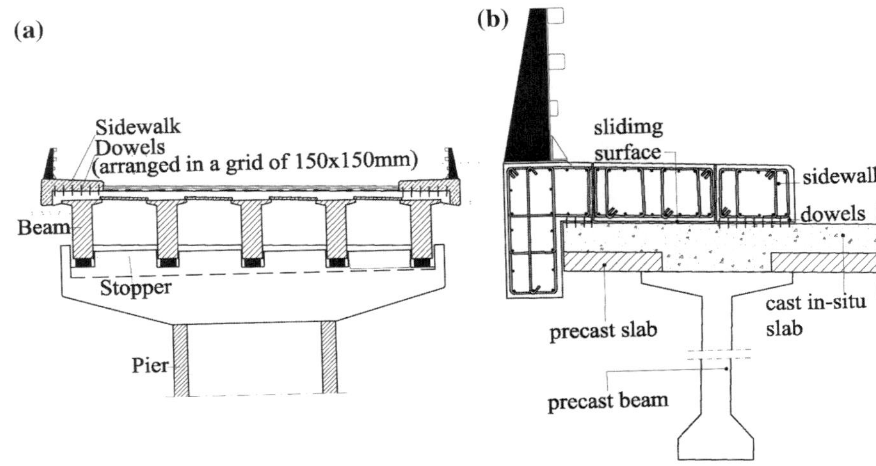

Fig. 3 **a** Cross-section of the sidewalk-restrainers at the anchorage length (Section A-A in Fig. 2), **b** cross-section of the sidewalks-restrainers (section B-B in Fig. 2)

The implementation of the proposed technique requires the proper anchoring of the sidewalks ends. More specifically, the inner end of the sidewalks (toward the centre of the bridge) is anchored to the bridge deck through dowels arranged in a grid of 150 × 150 mm, Fig. 3a. The anchorage length is about 2.0 m. The outer end of the sidewalks (toward the embankment) is anchored to the top of the wing-wall.

3 Serviceability and Seismic Performance of the Sidewalks-Restrainers

During the contraction of the bridge deck the sidewalks behave as tension ties. The developed tensile forces are balanced by the longitudinal reinforcement of the sidewalks. In Fig. 4 the idealized behaviour of a reinforced concrete tie is presented [13]. According to this figure the contribution of the concrete may be considered to increase the stiffness of the tensile reinforcement (tension stiffening effect).

The minimum design length of the segment of the sidewalks that is engaged in resisting the seismic effects in the longitudinal direction is determined by the serviceability requirements of the bridge. More specifically its calculation is based on the crack control according to the Code's requirements [14]. In this calculation the modulus of elasticity of the sidewalk-tie is taken 20 % greater than the modulus of elasticity of the steel according to Fig. 4. At this point, it is also necessary to note that, there is a difference between the requirements of the serviceability and

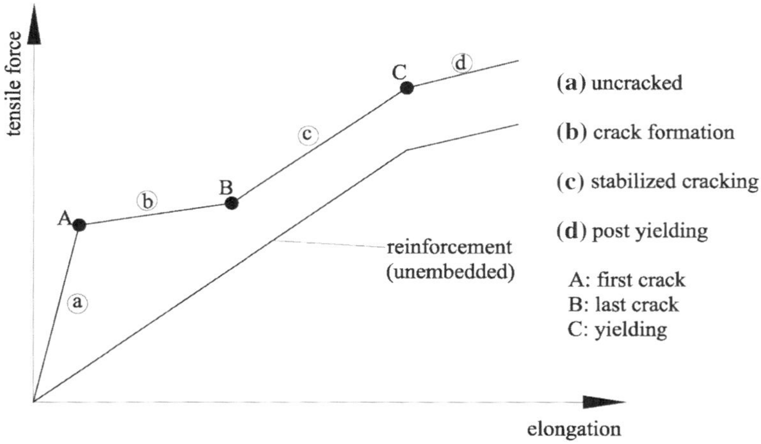

Fig. 4 Idealized behaviour of a reinforced concrete tie

seismicity needs regarding the longitudinal reinforcement ratio of the sidewalks. This happens because an increase in the number of bars, although it definitely improves the seismic resistance, it leads to additional secondary moments for the service limit state.

The maximum in-service change in the sidewalk's length Δl is derived by Eq. 1. The first part of Eq. 1 corresponds to the maximum in-service movement of the bridge. The deck contracts due to the maximum thermal contraction [15]. The second part of Eq. 1 corresponds to the change in the sidewalk's length due to the thermal contraction, creep and shrinkage. $\Delta T_{N,tot}$ is the sum of the maximum variation of the uniform bridge temperature contraction component $\Delta T_{N,con}$ and the equivalent uniform contraction temperature $\Delta T_{N,per}$ due to creep and shrinkage [16]. The first part of $\Delta T_{N,tot}$ was considered to be equal to 25 °C [15] while the second part was also considered equal to 25 °C. a is the coefficient of thermal expansion, assumed to be equal to 10^{-5} m/°C [14], L_{tot} is the total length of the continuous deck of the bridge and l_{eff} is the effective length of the sidewalk- restrainer.

$$\Delta l = \alpha \cdot \Delta T_{N,con} \cdot \left(\frac{L_{tot}}{2} - l_{eff} \right) + \alpha \cdot \Delta T_{N,tot} \cdot l_{eff} \quad (1)$$

The minimum length, l_{eff}, is derived by Eq. 2 as it is the only unknown factor. The first part of the equality corresponds to the deformation of the sidewalk. Δl is the maximum in-service change in the sidewalk's length and includes the unknown length l_{eff}.

$$\frac{\Delta l}{l_{eff}} \simeq 0.15\ \% \quad (2)$$

The deck's contraction due to creep, shrinkage and thermal effects causes tension of the sidewalks-restrainers. The developed axial force acting at the centre of the sidewalk's cross section at the anchorage joint causes a bending moment which compresses the bottom surface and has favourable effect on the distress of the deck due to the dead and traffic loads. On the other hand, the thermal expansion of the deck does not cause additional forces as the implied buckling of the sidewalk contribute to the elimination of the negative effects. The maximum lift of the sidewalks-restrainers due to the aforementioned inappreciable bend is of the order of some millimetres.

During the longitudinal earthquake the sidewalks-restrainers behave as tension-ties which restrain the longitudinal movements of the bridge deck. The forces of the sidewalks-restrainers are transferred to the wing-walls. Equation 3 gives the maximum tensile resistance F_{yd} of the sidewalks. In this equation A_s is the longitudinal reinforcement area and f_{yd} is the design yield stress of the steel bars (taking into account a safety factor 1.15) [14].

$$F_{yd} = A_s \cdot f_{yd} \tag{3}$$

The longitudinal reinforcement of the sidewalks-restrainers consists of bars of 14 mm or 16 mm diameter corresponding to a ratio equal to 2–4 % while dense transverse reinforcements of diameter 8–10 mm are used for the confinement of the cross-section.

4 Description of the Bridge Used as the "Reference" Case

The efficiency of the proposed restraining system was assessed by applying it on a ductile alteration of a precast I-beam bridge located at Asprovalta territory of Egnatia Odos Motorway and is given in Fig. 5a. This "reference" bridge has four spans and a total length equal to 137.6 m. The deck, Fig. 5b, consists of five simply supported precast and prestressed I-beams, precast deck slabs and a cast

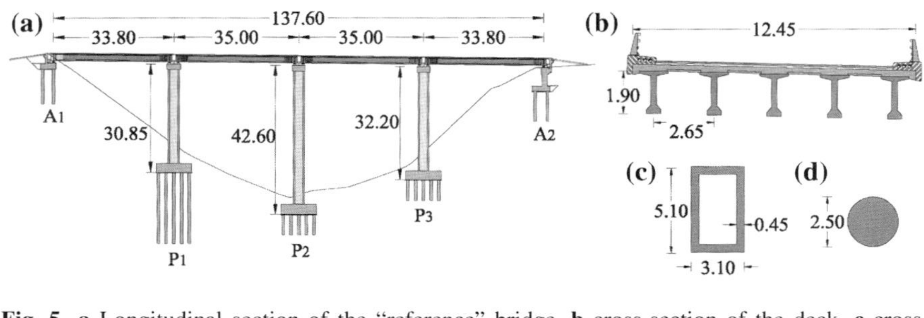

Fig. 5 **a** Longitudinal section of the "reference" bridge, **b** cross-section of the deck, **c** cross-section of the piers of the "reference" bridge and **d** cross-section of the piers of the ductile alteration of the "reference" bridge

in situ part of the slab. The deck is supported on both abutments and on the piers through low damping rubber bearings. The piers of the "reference" bridge, Fig. 5c, are hollow rectangular sections with external dimensions 3.1 × 5.1 m and a web thickness equal to 0.45 m. The bridge is founded on a ground type B according to the Greek seismic design code [17]. The design ground acceleration was equal to 0.16 g. The importance factor adopted was equal to $\gamma I = 1.3$ [18], while the behaviour factors were equal to 1.0 for the three directions.

The properties of the ductile bridge alteration used for the investigation on the efficiency of the proposed restraining system are the same with the initial one apart from the cross-section of the piers (circular cross-sections), Fig. 5d. This bridge has also active longitudinal seismic stoppers at the head of piers P_2 and P_3. The method used for the conversion of the "reference" bridge to a ductile system has analytically been presented in a previous study [19].

This bridge which is considered to be the "reference" case of the study is analysed considering ground accelerations 0.16 and 0.24 g.

The longitudinal reinforcement of the sidewalks corresponds to a ratio equal to 2 %. The minimum design length of the seismically active parts of the sidewalks was determined according to the serviceability requirements of the bridge. This length is equal to 13.8 m. The seismic active length of the sidewalks-restrainers must be greater or equal than the aforementioned minimum length. In this study the seismically active parts of the sidewalks is selected to be anchored at the first and third pier. Considering that the length of the end spans of the bridge is 33.8 m the total length of each sidewalk-tie is 36.8 m. This length is the span's length plus the anchorage length of the sidewalk into the wing-wall (5 m) minus the anchorage length of the sidewalk on the deck (2 m). Each sidewalk has a width equal to 1.0 m which is divided to three zones. The outer zones ($b_1 = 0.30$ m) are anchored to the deck while the central one ($b_2 = 0.40$ m) has the ability to slip on the bridge deck.

5 Evaluation on the Seismic Performance of the Proposed Restraining System

The seismic efficiency of the proposed restraining system was assessed using nonlinear time history analysis. Five artificial records compatible with the Eurocode 8- Part 1 [20] elastic spectrum and corresponded to 0.16 g and 0.24 g peak ground accelerations were used. The Newmark integration method was used, with time step $\Delta t = 0.01$ s and a total of 2000 steps (20 s of input).

Figure 6 illustrates the model of the modified bridge, in which the sidewalks are used as seismic restrainers. The superstructure was modelled by frame elements and the existing bearings by springs. The sidewalks-restrainers were modelled by springs, which act only in tension. Analysis was carried out by using the SAP 2000 program [21]. The plastic hinging in the piers was modelled by considering

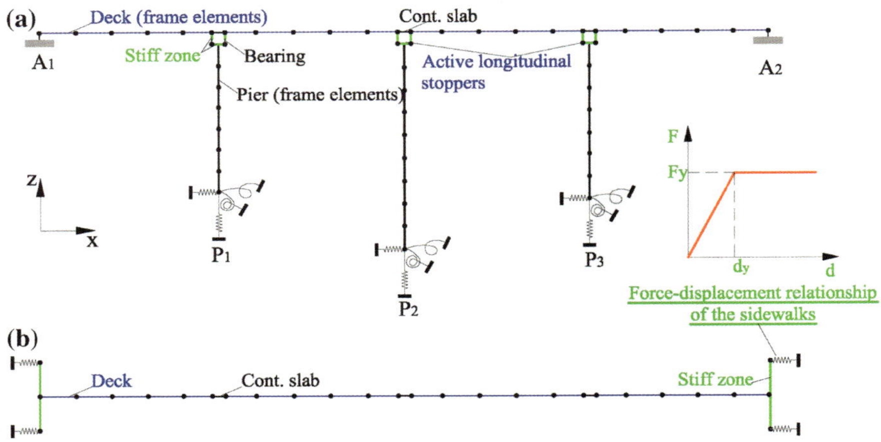

Fig. 6 The model of the upgraded bridge: a longitudinal section and b plan view

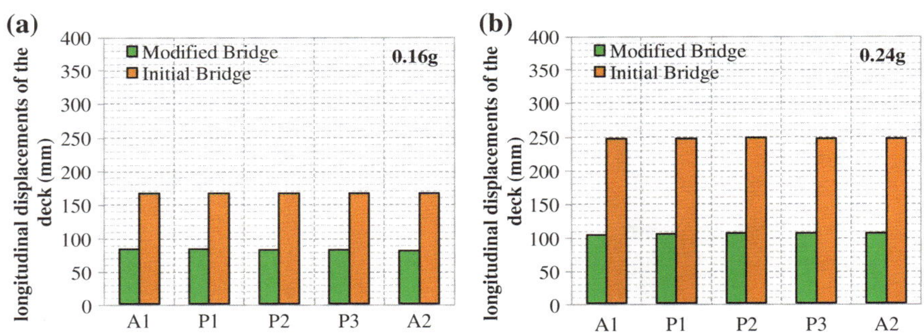

Fig. 7 The longitudinal displacements of the deck over the piers **a** ag = 0.16 g and **b** ag = 0.24 g

nonlinear rotational spring elements at the ends of the piers. The input parameters of the moment-rotation (M-θ) relationship were determined by fibre analysis performed in the computer program RCCOLA-90 [22] for each pier cross-section.

The analysis presented in the following paragraphs focuses on the longitudinal response of the bridge, as the method presented in this study deals with this direction of the bridge, which, as it is known, is more demanding than the transverse one. The response of the bridge to the transverse earthquake is more easily treated due to the presence of seismically active stoppers at the head of the piers.

The efficiency of the proposed restraining system was assessed by comparing the longitudinal movements of the modified bridge deck with the movements of the initial one. Figure 7 shows the longitudinal movements of the initial and the modified bridge systems for two ground accelerations 0.16 and 0.24 g. It is deduced that the longitudinal movements of the modified bridge are about 50 % smaller than the movements of the initial one for ground acceleration 0.16 g. The aforementioned percentage is increased for the case of ground acceleration 0.24 g to 58 %.

6 Conclusions

In this study the investigation on the seismic efficiency of an innovative restraining system for the improvement of the earthquake resistance of precast I-beam bridge was presented. The proposed methodology is based on the conversion of the end parts of the bridge sidewalks to seismic restrainers. Each sidewalk is divided to three zones. The outer zones are anchored to the deck while the central one has the ability to slip on the bridge deck. The one end of the sidewalk is anchored at the deck through high strength dowels, while the second is anchored at the top of the wing-wall. The efficiency of the proposed method is investigated by utilizing a precast I-beam bridge of Egnatia Odos Motorway in Greece. It is desirable to underline the following as the main results of the investigation:

- The proposed restraining system is indicated for precast I-beam bridges. It aims at the effective reduction of the displacements due to the longitudinal earthquake.
- The proposed methodology comprises an indirect technique for the improvement of the earthquake resistance of bridges and it does not affect the aesthetics of the bridge.
- The proposed earthquake resistant system deals effectively with the bridge's serviceability demands through the free slippage of the sidewalks on the main structure. Therefore, no buckling problems arise.
- The reduction in the maximum seismic displacements of the modified bridge system is significant. The resulting displacement magnitude is limited to 50 % of the displacement magnitude of the initial bridge system. The significant reduction in the deck's longitudinal movement also reduces the demand for voluminous bearings, whose thickness is mainly determined by the seismic displacements.
- From the economic point of view, the proposed restraining system has minor cost considering its effectiveness on the longitudinal earthquake, which is generally more difficult to be treated than the transverse one.

References

1. Priestley M, Seible F, Calvi G (1996) Seismic design and retrofit of bridges. John Wiley and Sons Inc, New York
2. Kelly TE (2001) Base isolation of structures design guidelines. S.E. Holmes Consulting Group Ltd, Auckland
3. Kelly TE (2001) In structure dumping and energy dissipation-design guidelines. S.E. Holmes Consulting Group Ltd

4. Naeim F, Kelly JM (1999) Design of seismic isolated structures, from theory to practice. John Wiley and Sons, Inc, New York
5. Ruangrassamee A, Kawashima K (2003) Control of nonlinear bridge response with pounding effect by variable dampers. Eng Struct 25(5):593–606
6. CEN (2003) Eurocode 8: design of structures for earthquake resistance, Part 2: Bridges
7. Tegou SD, Mitoulis SA, Tegos IA (2010) An unconventional earthquake resistant abutment with transversely directed R/C walls. Eng Struct 32(11):3801–3816
8. Mitoulis SA, Tegos IA (2010) Restrain of a seismically isolated bridge by external stoppers. Bull Earthq Eng 8(4):973–993
9. Caner A, Dogan E, Zia P (2002) Seismic performance of multi simple span bridges retrofitted with link slabs. ASCE J Bridge Eng 7(2):85–93
10. Sevgili G, Caner A (2009) Improved seismic response of multisimple-span skewed bridges retrofitted with link slabs. J Bridge Eng 14(6):452–459
11. Mikami T, Unjoh S, Kondoh M (2003) The effect of abutments as displacement limiting measure on seismic performance of bridges, www.pwri.go.jp/eng/ujnr/tc/g/pdf/19/4-1mikami.pdf, Accessed 5 Apr 2014
12. Nutt RV and Mayes RL (2000) Comparison of typical bridge columns seismically designed with and without abutment participation using aashto division i-a and proposed AASHTO LRFD provisions, Task F3-1(a)
13. CEB-FIB (1991) Model code 1990, final draft, CEB Bulletin No 204, 1991
14. CEN (2004) Eurocode 2: Design of concrete structures-Part 1: General rules and rules for buildings
15. CEN (2003) Eurocode 1: Actions on structures—Part 1-5: general actions—thermal actions
16. PCI (2014) Precast, prestressed concrete bridges, the high performance solution. comprehensive bridge design manual. Available via DIALOG: www.pci.org/publications/bridge accessed 12 Apr 2014
17. Ministry of public works of Greece (2003) Greek seismic code-EAK2000, Athens
18. Ministry of public works of Greece (2007) Guidelines for seismic design of bridges in combination with DIN-FB 102 103, 104, Athens
19. Tegos IA, Tegou SD, Markogiannnaki OG (2012) Seismic design of precast i-beam bridges based on ductility. Struct Eng Int 23(2):176–186
20. CEN (2003) Eurocode 8: Design of structures for earthquake resistance, Part 1: general rules, seismic actions and rules for buildings
21. Computers and Structures Inc (2007) SAP 2000 nonlinear, user's Reference manual, California
22. Kappos AJ (2002) RCCOLA-90: a microcomputer program for the analysis of the inelastic response of reinforced concrete sections. Department of Civil Engineering Aristotle University of Thessaloniki, Greece

Topographic Effect on Seismic Response of a High-Pier Bridge Subjected to Oblique Incidence Waves

Yin Gu, Chung C. Fu and M.S. Aggour

Abstract Oblique incidence waves play a vital role in the spatial variation of seismic ground motions. It is important to consider the oblique incidence of seismic wave in a valley or in other complex topographies. The numerical simulation method of free field with uniform topographic under oblique incidence waves is studied first. The accuracy is then tested by analytical solutions. A finite element model of bridge-soil system is built to analyze the seismic response of a continuous rigid-frame bridge with high piers located in a deep valley with the consideration of oblique incidence waves. The topographic effect on amplitude and distribution of internal forces where determined for different oblique incident angles and different shear wave velocities of the soil. The dynamic response of this bridge under P-wave and SV-wave are summarized. The results show that the angle of the oblique incidence waves has greater effect on internal forces of the piers and piles in deep valley than in uniform field shapes. If the bridge is located in a valley, the effect of oblique incidence waves on the bridge should be considered.

1 Introduction

It is a reasonable assumption that a small structure on a plain topography subjected to a uniform seismic wave is used in the analysis of a soil-structure dynamic interaction system. However, waves propagates in different ways in valley and

Y. Gu (✉) · C.C. Fu · M.S. Aggour
Department of Civil Engineering, University of Maryland, College Park, MD 20742, USA
e-mail: cinoa@fzu.edu.cn

C.C. Fu
e-mail: ccfu@umd.edu

Y. Gu
College of Civil Engineering, Fuzhou University, Fuzhou 350108, Fujian, China

A. Caner et al. (eds.), *Developments in International Bridge Engineering*,
Springer Tracts on Transportation and Traffic 9,
DOI 10.1007/978-3-319-19785-2_14

plain field [1]. The seismic wave propagation will be difficult to calculate owing to the complex topography, which has great effects on bridges with high piers, especially those with quite varied pier heights. This problem brought concerns to researchers [2, 3] that the seismic ground motion is obviously different in a valley than in the plain. These differences and their spatial effects should be considered. The oblique incidence of the waves is one of the most important spatial effect. Based on common models, earthquakes are mainly expected as shear waves propagating vertically. However, in near fields, the seismic waves has oblique incidence with spatial distributions characters. It means that there are different amplitudes and durations at different locations of the structure. Two directions including horizontal and vertical directions coupled together has great effect on the soil-structure dynamic interaction system. In recent years, there are many studies for seismic response considering oblique incidence waves. But most of the studies concern about the analysis of underground soil-structure systems, which include subway stations and buried pipes. The studies have shown that the response of the underground structure under oblique incidence waves is different from those under vertical waves. The structure is not safe by neglecting the oblique incidence waves under this condition [4–7]. Bogdanoff found that the non-uniform excitation is significantly different in various types of structures [8]. The time-lag effect in seismic wave propagation has an influence on large-span structure. The spatial distribution in seismic field was first considered in the European code [9]. The spatial distribution has also been included in seismic analysis of large-span bridges in Chinese code [10]. Liu studied the seismic response of bridges under oblique incidence and the topographic effect [11]. The results show that the response of the structure is larger in comparison with the traditional consideration of vertical wave. So the effect of oblique incidence waves on the bridge should be considered for complex topographic. This paper proposed a method for how to input the seismic waves considering its oblique incidence in finite element analysis. The method was then used in a case study to show the effect of the topography and the angle of the oblique incidence waves.

2 Equivalent Artificial Boundary Element and Its Seismic Input

Finite element method is widely used in numerical analysis considering local non-regular topography. Studies shows that the damping ratio of the structure—soil system increases by 25 % compared with the model with spring and dampers, which response are different from the former [12, 13]. The system including the structure and the surrounding soil is more accurate. The limited computing area should be cut from the unlimited earth medium. Artificial boundary is added to the boundary to simulate the radial damping of continuous medium. It should be verified that the radial waves do not reflect back to the limited computing area.

An artificial boundary element proposed is used here. The method is to extend a layer of elements with the same type of the inside elements along the normal direction. The outside boundaries of it are fixed. The accuracy of equivalent element has been verified [14]. The equivalent shear modules, elastic modules and equivalent damping are

$$\begin{cases} \tilde{G} = hK_{BT} = 2\alpha_T h\dfrac{G}{R} \\ \tilde{E} = 2\dfrac{(1+\tilde{v})(1-2\tilde{v})}{(1-\tilde{v})}hK_{BN} = 2\alpha_N h\dfrac{G}{R}\cdot\dfrac{(1+\tilde{v})(1-2\tilde{v})}{(1-\tilde{v})} \end{cases} \quad (1)$$

$$\tilde{\eta} = \frac{\rho R}{3G}\left(2\frac{c_s}{\alpha_T} + \frac{c_p}{\alpha_N}\right) \quad (2)$$

in which R is the distance from wave sources to artificial boundary; h is thickness of the equivalent element. c_s and c_p are velocities of S and P waves, respectively; G is the shear modulus of medium; ρ is the mass density of medium; α_T and α_N are parameters. K_{BT} and K_{BN}, usually used in viscoelastic artificial boundaries, can be calculated in Eq. (1).

After setting the artificial boundary, the waves will be input by equivalent loading, which cause the same displacement, velocity and stress as the free wave field. The equation of the equivalent load input at a point B at the base is

$$\begin{aligned} P_{BN}(x_B, y_B, t) &= \left[\sigma(x_B, y_B, t) + C_{BN}\dot{u}(x_B, y_B, t) + K_{BN}u(x_B, y_B, t)\right]\sum A_i \\ P_{BT}(x_B, y_B, t) &= \left[\tau(x_B, y_B, t) + C_{BT}\dot{u}(x_B, y_B, t) + K_{BT}u(x_B, y_B, t)\right]\sum A_i \end{aligned} \quad (3)$$

in which P_{BN} and P_{BT} are equivalent loads of the normal and tangential directions at the node B; C_{BN} and C_{BT} equals to $\rho Rc_s/G\alpha_T$ and $\rho Rc_p/G\alpha_N$; $\sigma(x_B, y_B, t)$ are the normal stresses of node B; $\tau(x_B, y_B, t)$ is tangential stress of node B; $u(x_B, y_B, t)$ and $v(x_B, y_B, t)$ are displacement responses of node B in free field; $\dot{u}(x_B, y_B, t)$ and $\dot{v}(x_B, y_B, t)$ are velocity responses of node B free field. The equivalent loads can be calculated according to Eq. (3), which needs to obtain the response of the free field.

3 Input Seismic Waves with Oblique Incidence

3.1 Numerical Method

P wave and SV wave can be interchanged when they pass through the boundary of different medium. Waves of oblique incidence produce P and SV waves and their reflected waves shown in Fig. 1.

Fig. 1 Sketch of oblique incidence propagation

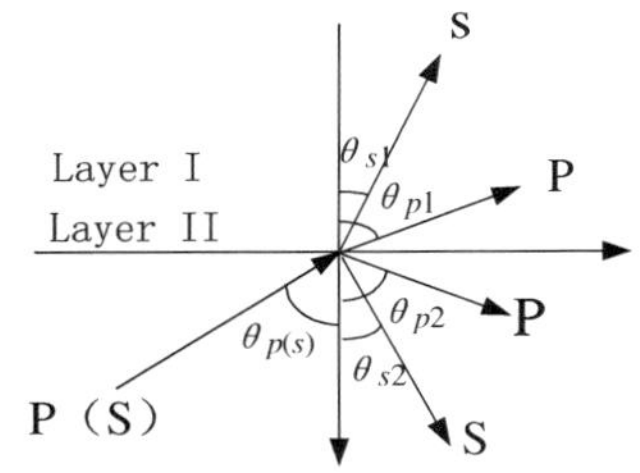

Let the horizontal direction as x axis; vertical direction as y axis. There is an angle, θ, between motion direction and y axis. The apparent wave velocity, propagating as elastic wave with velocity c, along horizontal direction is shown in Eq. (4),

$$c_x = \frac{c_{p1}}{\sin \theta_{pi}} \tag{4}$$

in which c_{p1} is the propagation velocity in medium of layer one. c_x is the horizontal apparent wave velocity. According to Snell rule, the shear velocity of horizontal direction in each soil layer is the same. The horizontal velocities of seismic waves have the following relation,

$$\frac{c_{s1}}{\sin \theta_{s1}} = \frac{c_{p1}}{\sin \theta_{p1}} = \frac{c_{s2}}{\sin \theta_{s2}} = \frac{c_{p2}}{\sin \theta_{p2}} = c_x \tag{5}$$

in which c_{s1} is the velocity of S wave propagating in layer one. c_{p1} is the velocity of P wave propagating in layer one. c_{s2} is the velocity of S wave propagating in layer two. c_{p2} is the velocity of P wave propagating in layer two.

Considering the effect of traveling wave, when P wave propagates in oblique incidence, the displacement in free field can be written in Eq. (6) in elastic medium of the layer.

$$u(x, y, t) = u\left(0, y, t - \frac{x}{c_x}\right) \tag{6}$$

Substitute t as Δt (time step), the equation can be written as

$$u(jc_x \Delta t, y, t) = u(0, y, t - j\Delta t) \tag{7}$$

The free fields in half space of response under waves of oblique incidence can be calculated by method of time history method of one dimension [15]. According to the wave motion in Eq. (7), displacements of left, right and bottom boundaries can be calculated. The finite element method of lumping mass, which means motion of any node is only related to the node of neighbor joint, is used in Eq. (3) to calculate the time-history force in the three faces of the boundaries. Considering the accuracy and stable of the numerical solution, this method should meet the following condition: $\Delta t <= \frac{\Delta y}{c_s}$ in S wave and $\Delta t <= \frac{\Delta y}{c_p}$ in P wave.

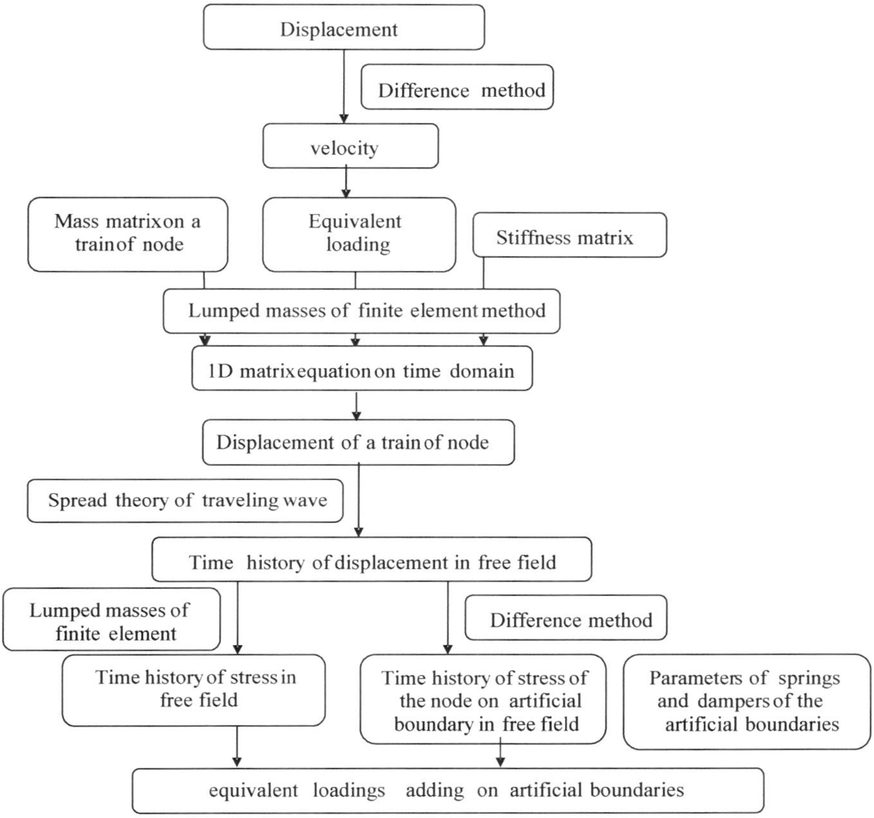

Fig. 2 Flowchart of solution for the equivalent load on node of artificial boundary

According to the method mentioned above, the problem of the seismic wave propagation is transferred to the problem of wave source. A program of adding equivalent loading on the artificial boundaries is made by MATLAB software. The procedure is shown in Fig. 2.

3.2 Verification of the Procedure

A model of uniform layered soil, as the one used in reference [15], is selected for testing the accuracy of the method. A layer of artificial boundaries is set on the left, right and bottom sides of the free field. The equivalent loading is added on each node of the boundary both in x and y directions. Sketch of the overlay is shown in Fig. 3a. The displacement time history of input curve for point B is shown in Fig. 3b. The parameters of the soil are shown in Table 1.

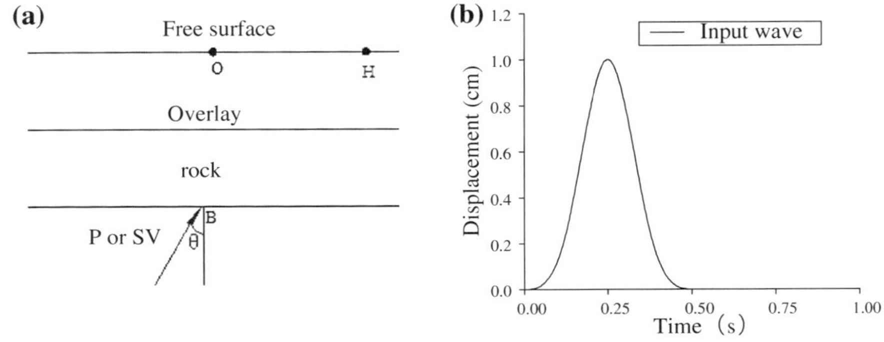

Fig. 3 Sketch of the overlay model and the input wave. **a** Location of the input wave. **b** Time history curve of input wave

Table 1 Parameters of the soil

Type	Thickness (m)	Density(kg/m^3)	V_s(m/s)	V_p(m/s)
Overlay	50	1000	200	346
Rock	50	1500	500	866

The results shown in Fig. 4 are close to the results in reference [15], which validates the input method proposed.

4 Application to a Selected Case Study

4.1 The Case Study

A numerical model was built of a continuous rigid bridge with the total length of 560 m and its spans of 100 + 180 + 180 + 100 m. The pier is a RC double wall of 2 m thickness, 9.0 m depth, and about 30 m height. The diameter of the circle pile is 2.8 m. The main features and the basic dimensions of the bridge are shown in Fig. 5a. The numerical model of the bridge with special topographic shape is shown in Fig. 5b. The parameters of the model elements and soil are shown in Tables 2 and 3 respectively.

4.2 Input Seismic Wave

We selected the Anza Earthquake record that has a short duration to reduce the time for computing. The time history curves of acceleration and displacement are shown in Fig. 6. The peak of the displacement is 0.16 cm. The time step is 0.004 s.

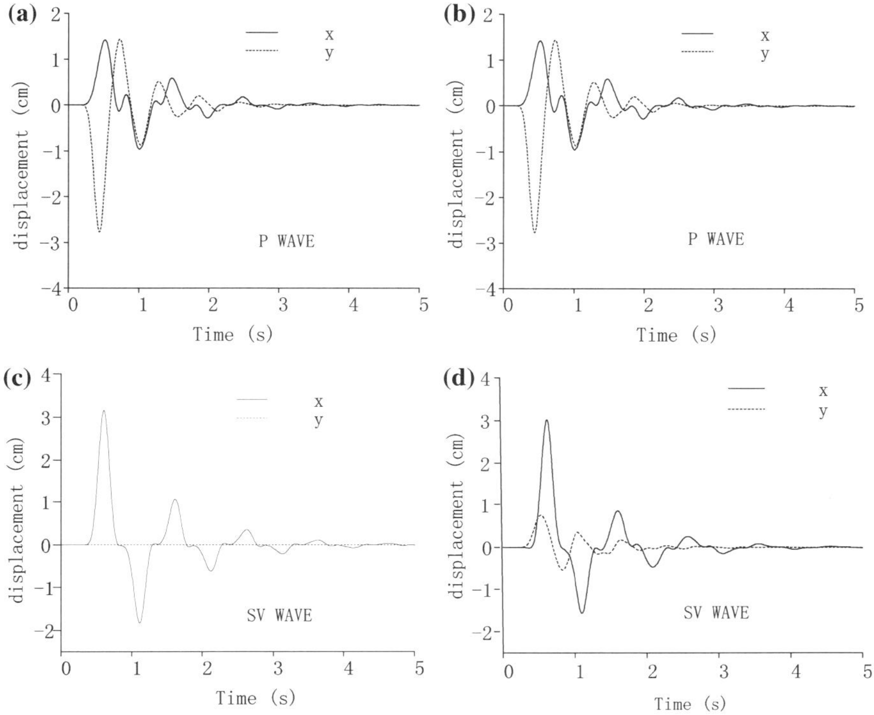

Fig. 4 Time history curves of displacement with different angles: **a** $\theta = 0.001°$ P wave, **b** $\theta = 30°$ P wave, **c** $\theta = 0.001°$ SV wave **d** $\theta = 30°$ SV wave

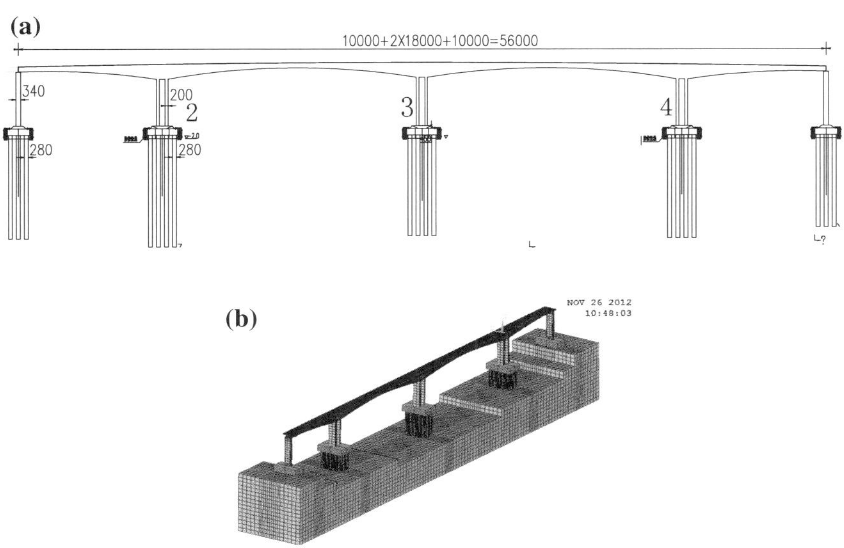

Fig. 5 Layout of the long-span bridge and FEA model: **a** Elevation **b** Soil-bridge finite element model

Table 2 FEA model and element

Element	BEAM188	MASS21	SOLID45	BEAM4
structure	Girder, pier, pile	Lump mass	Soil	Rigid

Table 3 Material parameters of soil

V_S (m/s)	V_P (m/s)	μ	P (kg/m^3)	Thickness of soil (m)
500	866	0.25	2200	30

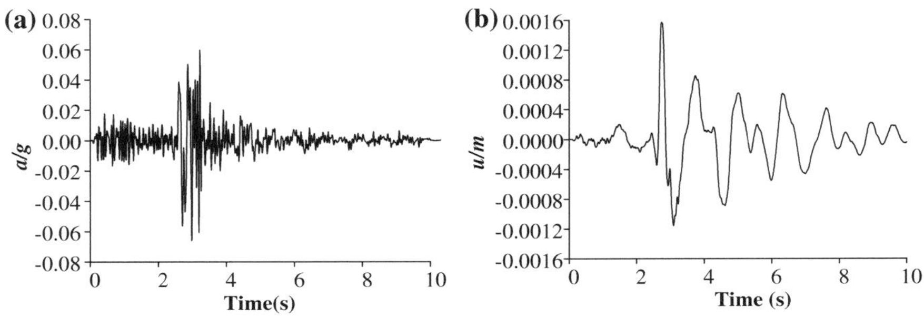

Fig. 6 Input earthquake wave: **a** Acceleration, **b** Displacement

4.3 Results and Discussion

The results of the internal forces are compared by inputting the seismic wave in vertical direction and on an oblique incidence. The bottom and the top of the pier are the most vulnerable locations in this rigid continuous bridge. The force at the bottom is larger than that at the top in this case. So the internal forces at the top of piers are studied. The location of the numbered piers are shown in Fig. 5a. The amplitudes of the internal forces of piers subjected to P waves of different input angles of the oblique incidence wave are shown in Fig. 7. The results showed that the shear force of the middle pier is smaller than the other two piers. The shear force and moment of the piers with seismic wave in the vertical direction are smaller than the force and moment with inputting wave at a 30° angle. Conversely, the axial force of the piers is larger in the vertical direction.

The amplitudes of the internal forces of piers with SV waves of different input angles of the oblique incidence waves are shown in Fig. 8. The shear force of the middle pier is smaller than those of the other two piers also. But the results of the SV waves showed a reverse trend from the P waves. The shear force and moment of the piers with seismic wave in the vertical direction is larger than the force and moment with inputting wave at a 30° angle. Conversely, the axial force of the piers is smaller in the vertical direction.

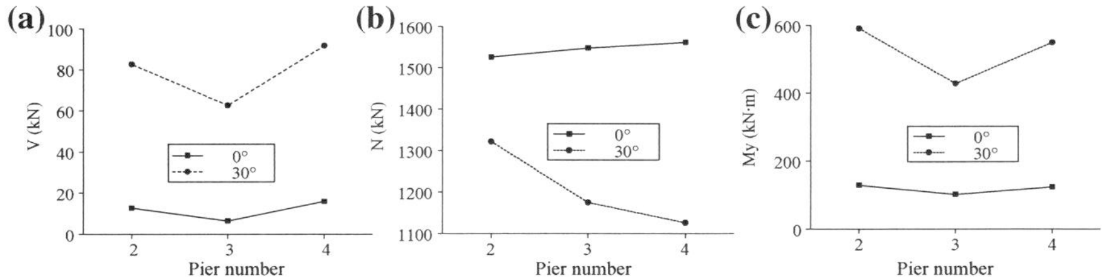

Fig. 7 Maximum amplitude of internal forces at bottom of piers under oblique incidence P waves: **a** Shear force V, **b** Axial force N, **c** Moment My

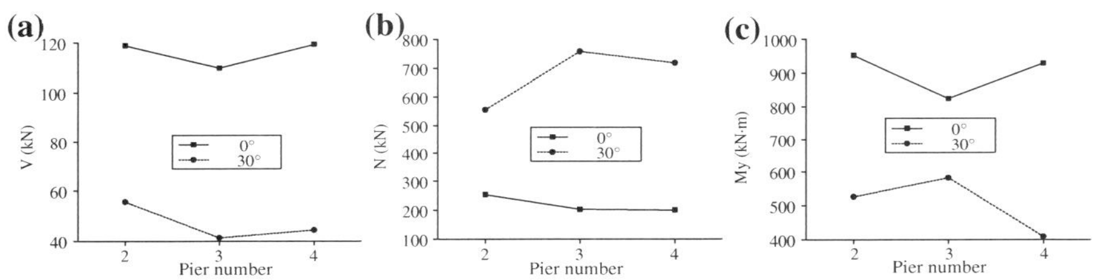

Fig. 8 Maximum amplitude of internal forces at bottom of piers under oblique incidence SV waves: **a** Shear force V, **b** Axial force N, **c** Moment My

Considering ratio of the values computed from vertical and oblique incidence input, an index can be calculated as follows:

$$\eta = \frac{\max|F| - \max|F_0|}{\max|F_0|} \tag{8}$$

in which F_0 is the internal force of the piers including shear forces, axial forces and moment computed by vertical input; F is the internal force computed by oblique incidence wave.

The indices of non-uniform effects for piers in valley under oblique incidence waves at 30° are shown in Table 4. It is shown that the oblique incidence waves have great effect on the piers of bridges in valley.

Table 4 Indices of non-uniform effects for piers of valley under oblique incidence waves at 30°

Internal forces	η_P			η_{SV}		
	Pier 2 (%)	Pier 3 (%)	Pier 4	Pier 2 (%)	Pier 3 (%)	Pier 4 (%)
V	552	874	477	−53.28	−62.39	−62.80
N	−13	−24	−27	118.86	274.26	259.41
My	359	319	344	−44.76	−29.41	−56.02

5 Conclusions

This study is focused on considering the effect of oblique incidence waves and on building a model of bridge-soil dynamic interaction. The results showed that the shear force and moment of the piers under P wave in the vertical direction is smaller than the force and moment with inputting wave at a 30° angle. While the axial force of the piers is larger in the vertical direction. For the seismic SV waves, on the contrary, the shear force and moment of the piers under wave in the vertical direction is larger than the force and moment with inputting wave at a 30° angle. The axial force of the piers is smaller in the vertical direction. It is shown that the oblique incidence has great effects on the piers of the bridges in valley.

References

1. Trifunac MD (1973) Scattering of plane SH waves by a semi-cylindrical canyon. Earthquake Eng Struct Dyn 1:267
2. Liu BD, Zhou ZhH, Liu PX, Li XJ, Wang W (2011) Influence of V-shaped canyon site on ground motions for incident SV waves. Earthquake Eng Eng Vibr 31(2):17–25 (In Chinese)
3. Zhou GL (2010) Canyon topography effects on seismic response of multi-support bridges. Institute of Engineering Mechanics, China Earthquake Administration, Harbin (In Chinese)
4. Du XL, Chen W, Li L, Li LY (2007) Preliminary study of time-domain seismic response for underground structures to obliquely incident seismic waves. Technol Earthquake Disaster Prev 2(3):290–296
5. Wang Q, Chen J, Li J (2008) Seismic response analysis of underground pipeline under obliquely incident seismic waves. J Huazhong University of Science and Technol 25(4):283–286
6. Chen W. (2007). Seismic response analysis of underground structures under obliquely incident seismic waves. Master dissertation, Beijing University of Technology, Beijing, China. (In Chinese)
7. Nazmy Aly S, Abdel-Ghaffar Ahmed M (1992) Effect of ground motion spatial variability on the response of cable-stayed bridges. Earthquake Eng Struct Dyn 21:1–20
8. Bonganoff JL, Goldberg JE, Schiff AJ (1965) The effect of ground transmission time on the response of long structures. Bull Seism Soc Am 55:627–640
9. EN (1995) Eurocode 8: Structures in seismic regions design, Brussels European Committee for Standardization
10. JTG/TB 02-01-2008 (2008) The rules of seismic design for highway bridges. The Department of Transportation. People's Communication Press, Beijing, China. (In Chinese)
11. Liu L (2005) Random vibration analysis of bridges considering topographic effects. Ph.D. dissertation, Beijing University of Technology, Beijing, China. (In Chinese)
12. Zhang, J., Nicos, M. (2001) Seismic Response Analysis of Highway Overcrossings Including Soil-Structure Interaction, PEER Report, University of California, Berkeley
13. Sun LM, Zhang ChN, Pan L (2002) Lumped-mass model and its parameters for dynamic analysis of bridge pier-pile-soil system. Chin J Tongji Univ 30(4):409–415
14. Gu Y, Liu JB, Du YX (2007) 3D Consistent viscous-spring artificial boundary and viscous-spring boundary element. Eng Mech 24(12):31–38 (In Chinese)
15. Wang, Y. (2007). Research on the Numerical Method for Asynchronous Seismic Wave Motions and its Application in Dynamic Analysis of Structures. Ph.D. dissertation, Tsinghua University, Beijing, China. (In Chinese)

Part IV
Condition Assessment and Structural Health Monitoring

Influence of the Primary Bridge Component Condition on the Overall Bridge Condition Rating

R. Hamid, Y. Khairullah and A.R. Khalim

Abstract Bridge condition rating is evaluated based on the material condition of the secondary and primary bridge components. This paper aims to investigate the influence of the condition of the primary bridge components to the overall condition of pre-stressed concrete beam bridge (PCBB), reinforced concrete beam bridge (RCBB) and steel beam bridge (SBB). Four primary bridge components namely surfacing, deck slab, beam/girder and abutments are used as input parameters and bridge condition rating as output parameters. This study utilizes multiple linear regression analysis (MRA) and artificial neural networks (ANN) to investigate the variance of the bridge condition rating with respect to the condition of the primary bridge components. The MRA results show that 62.83, 91.77 and 86.18 % of the proportion of the variance in the condition rating of PCBB, RCBB and SSB are explained by all the primary bridge components in the range of the training data set. Meanwhile ANN yields 67.35, 90.54 and 81.77 % for PCBB, RCBB and SSB, respectively. The results indicate that the condition rating of surfacing, deck slab, beam/girder and abutments highly contribute to the condition rating of RCBB and SSB, however for the PCBB, the influence is slightly lower. In term of modeling, MRA shows better performance for RCBB and SBB; however ANN seems suitable for PCBB.

1 Introduction

Generally, bridges can be constructed as single-span or multiple-span bridges with concrete or steel as its superstructure system. In highway networks of some countries, the majority of the bridge structures employ concrete (reinforced and

R. Hamid (✉) · Y. Khairullah · A.R. Khalim
Department of Civil and Structural Engineering, Faculty of Engineering and Built Environment, Universiti Kebangsaan Malaysia, Bangi, Malaysia
e-mail: roszilah@eng.ukm.my

A. Caner et al. (eds.), *Developments in International Bridge Engineering*, Springer Tracts on Transportation and Traffic 9,
DOI 10.1007/978-3-319-19785-2_15

pre-stressed) as the main constituent material [1, 2]. The Bridge Management System (BMS) of Malaysia maintains more than 7000 bridges along the federal highway. Majority of these bridge superstructures (about 80 %) utilise concrete and more than 9 % utilise steel. In term of structure system, simply supported bridges and culverts are the main structures of these bridges which are about 26 and 69 %, respectively [1].

BMSs of some countries classify different bridge components or bridge elements for their bridge condition rating systems. Many researchers utilised the classification based on the common practice in their home country in bridge assessment researches; for example, some researchers have divided bridges into eight components [3] or thirteen components [4]. Sasmal et al. [5] categorized the deck, superstructure and substructure as the main bridge components in their research and divided these components into 13, 16 and 20 elements, respectively. Chen [6] categorised and evaluated a beam bridge condition into four main bridge components namely substructure, bearing, beam and accessory structure. These four components are evaluated based on 20 assessment criteria that can be inspected by close visual inspections according to Chinese Bridge Maintain Codes. However, it is sometimes unrealistic to classify bridge components which comprise of the same materials or elements and expect them to deteriorate in the same manner.

For infrastructure condition rating systems, the inspector will generally use visual inspection as the first step of any condition assessment procedure [7–9] unless a structure cannot be visually assessed. Srinivasan and Parlikad [9] stated that one of the key challenges with the visual inspection is that different inspectors grade the degradation based on their perception. For example, one inspector might grade the degradation to be 6 while another can say it is rather 5. These condition rating systems which are based on visual inspection data then can be used for predicting the performance of individual bridges [10].

For BMSs optimization, there is a need to develop and integrate reliable and effective decision support models that include: condition rating models, deterioration prediction models, risk assessment models and maintenance optimization models [2]. In modelling of such models, a mathematical method such as an artificial neural network (ANN) and multiple regression analysis (MRA) can be useful to handle the uncertainty and subjective judgment of the bridge inspectors. Artificial Neural Networks (ANNs) have been used in many field areas to detect nonlinearities relationship between input vectors and the corresponding target vectors [11, 12]. A feed-forward artificial neural network (FFANN) is one of the ANN models that have been used widely in civil engineering computations such as bridge condition assessment [6, 13].

The purpose of this study is to investigate the influence of the condition of the primary bridge component to the overall condition of pre-stressed concrete beam bridge (PCBB), reinforcement concrete beam bridge (RCBB) and steel beam bridge (SBB). Four primary bridge components namely surfacing, deck slab, beam/girder and abutments are used as input parameters and bridge condition rating as output parameters. This study utilizes multiple linear regression analysis

(MRA) and artificial neural networks (ANN) to investigate the variance of the bridge condition rating in respect to its condition of primary bridge component.

2 Methods

2.1 Type of Bridge and Bridge Component

In the Annual Bridge Inspection Manual (ABIM) [14], the bridge structure systems consist of simply supported bridge; continuous span bridge; cantilever bridge; frame bridge; arches bridge; trusses bridge; bailey bridge, suspension bridge, cabled-stayed bridge and culverts. The bridge components of these bridges are classified as either primary or secondary elements. The primary components include surfacing (S), deck slab (D), beams or girders (B) and abutments (A), while the secondary components include parapet, expansion joints, bearings, slope protection and drainage. In terms of the construction material of the superstructure, these bridges are divided into concrete bridges, steel bridges and masonry bridges. In this study, the primary components of pre-stressed concrete beam bridge (PCBB), reinforced concrete beam bridge (RCBB) and steel beam bridge (SBB) are modelled in order to evaluate the entire bridge condition rating. The condition rating based on visual inspection of 29 PCCB, 40 RCBB and 36 SBB bridges are investigated in this study in order to model relationship between the primary component conditions to the entire condition of the bridges.

2.2 Bridge Condition Rating

The condition rating of a bridge structure is a process of assessing the overall condition and integrity of the structure as a whole, so that appropriate steps can be taken either for repair and rehabilitation or replacement. If condition ratings are available for a relatively long period of time, the data can be used to establish bridge condition rating and bridge deterioration models. Such models can then be used to estimate the status and the future condition of a bridge for the purpose of planning future bridge maintenance and/or repair activities [15].

The condition rating of an existing bridge structure involves a number of steps, which combines the different strength and serviceability parameters obtained from the structure and judgment by bridge's inspector. In most of the situations, the judgment cannot be estimated quantitatively, since the condition of the structure is described qualitatively. Furthermore, the structural condition rating of a structure depends on the type of structure and its intended service [16].

Condition rating of single span bridge is evaluated base on its component condition ratings which include primary components and secondary components as in ABIM [14]. Primary components include surfacing, deck slab, beams or girders

Table 1 The condition rating system based on severity of defect

Rating scale	General definition
1	No damage is found and no maintenance is required
2	Damage is detected and it is necessary to record the condition for observation purposes
3	Damage detected is slightly critical and thus it is necessary to implement routine maintenance work
4	Damage detected is critical and thus it is necessary to implement repair work or to carry out detailed inspection to determine whether any rehabilitation works are required or not
5	Being heavily and critically damaged and possibly affecting the safety or traffic, it is necessary to implement emergency temporary repair work immediately, rehabilitation work without delay after the provision of a load limitation traffic sign or replacement work

and abutments, meanwhile the secondary components include parapet, expansion joints, bearings, slope protection and drainage. The ratings are then assigned to the above components based on a scale of 1–5 to individual components by bridge inspectors using the guidelines established by the JKR as shown in Table 1.

2.3 Multiple Regression Analysis

Multiple regression analysis (MRA) is widely applied in modeling of bridge engineering problems whether in bridge components level or in entire bridge management systems [12, 17] due to its frequent use in predictions and the easy interpretation of the resulting models. In this study, MRA is used to analyze the relationship between the condition rating of primary bridge components and the entire bridge condition rating. In this case, multiple regression analysis addresses one output parameter, which is the bridge condition rating value, and four input parameters, which are the primary bridge conditions. If the bridge condition rating is y, primary bridge conditions are $x_1, x_2, x_3, \ldots, x_n$, then the model is given as follows:

$$y_i = \beta_0 + \beta_1 x_{1i} + \beta_2 x_{2i} + \beta_3 x_{3i} + \cdots + \beta_n x_{ni} + \varepsilon_i \tag{1}$$

where β is the coefficient of individual primary bridge conditions, n the number of primary bridge component conditions, $i = 1, 2, 3, \ldots, N$ is the number of data sets and ε represents the error of the model.

Two MRA models are explored in this study, namely MRA1, as shown by Eq. 1 and MRA2, which is a quadratic function as shown by Eq. 2.

$$y_i = \beta_0 + \beta_1 x_{1i}^2 + \beta_2 x_{2i}^2 + \beta_3 x_{3i}^2 + \cdots + \beta_n x_{ni}^2 + \varepsilon_i \tag{2}$$

2.4 Artificial Neural Network

An artificial neural network (ANN) is used to establish relationship between input data and output data within the domain of the training data set. ANN which consists of an input layer, a single of hidden layer and an output layer are used in this study used to establish relationship between primary bridge components and bridge condition rating. Four numbers of input parameters namely surfacing, deck slab, beam/girder and abutments were used in the input layers, and one parameter, the bridge condition rating, was used in the output layer. Figure 1 shows a typical ANN structure for modeling bridge condition ratings.

Various training algorithm is available in ANN such as scaled conjugate gradient back-propagation (*trainscg*), resilient back-propagation (*trainrp*), Levenberg-Marquardt algorithm (*trainlm*) and etc. The Levenberg-Marquardt algorithm (*trainlm*), which is suitable for small and medium size networks and can most rapidly be used to train a data set [18], was chosen for this study. Equation 3 shows the Levenberg-Marquardt method for bias and weight in an artificial neural network.

$$\Delta w(\text{or } \Delta b) = (JTJ + \mu I) - 1JTe \tag{3}$$

where Δw is weight change, Δb is bias change, J is the error for every weight (or bias) in a Jacobian matrix, μ is the scale, I is the identity matrix and e is vector error.

In neural network training processes, the errors are usually assessed with the sum squared error (SSE) and mean squared error (MSE). In each end cycle, SSE or MSE is calculated, whereas the weight and bias are updated based on the errors calculated.

In this study, a non-linear function of *tan-sigmoid* which is in the range [−1, +1], is applied in the mapping and learning process in hidden layers by reliable weight connections and activation functions. The non-linear activation functions are chosen for the neurons in the hidden layer, whereas for the neurons in the output layer, the linear activation function is used.

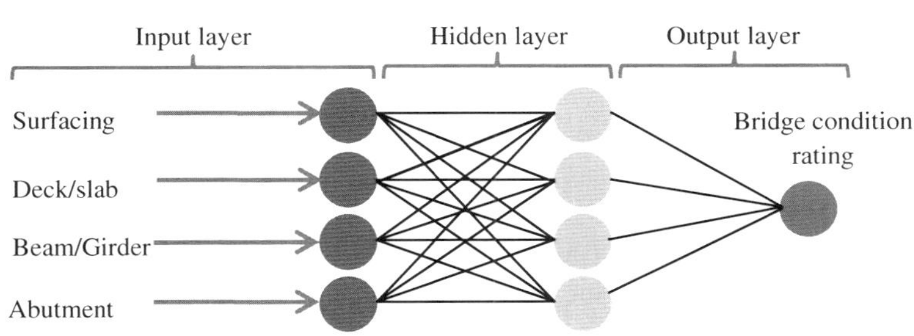

Fig. 1 Structure of typical ANN model

Training is the most important phase in designed neural networks. During the training process, the network will change the weight and bias for the learning input/output interrelationship. For every input data point that enters the network, the output generated by the network will be compared with the target. This algorithm will arrange or adjust the network parameters to minimize the mean square error.

In this study, numerous models were created by varying the number of neurons in hidden layer. Because there is no clear guideline for determining the number of neurons in the hidden layer, a trial and error approach is commonly adopted and was practiced in this study.

2.5 Model Evaluation

In evaluation of the developed models, mean square error (MSE), coefficient of determination (R^2) and Fisher's F Test is considered to evaluate the performance of the model. The F-statistics of the results of analysis of variance are commonly used to confirm the significance of the MRA model. A very low probability value for Fisher's F test demonstrates a very high significance for the regression model. These criteria have been widely used in the evaluation of mathematical models by previous researchers [12, 19]. These criteria are always related to error, consistency and accuracy. The evaluation of error, consistency and accuracy is concerned with the average of the difference between the target value and predicted value.

3 Results and Discussion

3.1 MRA Results

The best MRA models for PCBB, RCBB and SBB yield 62.83, 91.77 and 86.18 % variations in the condition rating of PCBB, RCBB and SSB bridges, as accounted for through the combined effects of the primary component condition rating variables indicated by R^2. In MRA modelling of PCBB condition rating, the quadratic form of multiple regressions seems suitable to explain the impact of primary component condition rating on condition rating of entire bridge. Meanwhile, the best model for RCBB and SSB were obtained from simple form of MRA. Table 2 shows the detailed results of multiple regression analysis for the condition rating model of all bridge superstructure configurations.

The verification results for the above models fit 76.56, 86.42 and 75.09 % of the testing data variation in the PCBB, RCBB and SBB bridges using MRA1. Meanwhile, MRA2 is able to fit 81.59, 85.64 and 74.91 % of the testing data variation in the PCBB, RCBB and SBB bridges. The best multiple regression analysis models relating bridge condition ratings to their explanatory primary component condition ratings are given in Table 3.

Table 2 Performance of MRA models for PCBB, RCBB and SBB bridges

Bridge superstructures	MRA1			MRA2		
	MSE	R^2 training	R^2 testing	MSE	R^2 training	R^2 testing
PCBB	0.1382	0.5963	0.7656	0.1272	0.6283	0.8159
RCBB	0.1098	0.9177	0.8642	0.1514	0.8866	0.8564
SBB	0.1331	0.8618	0.7509	0.1505	0.8438	0.7491

Table 3 The best MRA condition rating models for PCBB, RCBB and SBB bridges

Bridge superstructures	Best MRA models
PCBB	$1.3745 + 0.0164S^2 + 0.0282D^2 + 0.0535B + 0.0626A$
RCBB	$0.3861 + 0.0040S + 0.2747D + 0.2498B + 0.3712A$
SSB	$0.0506 + 0.0986S + 0.4360D + 0.3366B + 0.1397A$

Referring to the multiple regression analysis output, it shows that the overall equations listed in Table 3 are found to be statistically significant as indicated by (F = 7.19, Sig.-F < 0.000), (F = 69.73, Sig.-F < 0.000) and (F = 34.29, Sig.-F < 0.000) for PCBB, RCBB and SSB bridges, respectively. However, the relationship between the primary bridge component condition and the entire condition rating of PCBB bridge shows weak relationship compared with RCBB and SBB bridges.

3.2 ANN Results

Primary bridge components condition ratings were used as input parameters in network training of modelling three types of bridge superstructure systems. The condition ratings of these bridge components are rated based on material defects likely to be found on visual inspection using guideline as in ABIM [14]. Four primary bridge components namely: surfacing, deck slab, beams/girders and abutments are introduced as input parameters in the network modelling and entire condition rating of PCBB, RCBB and SBB bridge as output parameter. The training process of the networks is evaluated based on plotting the Mean Squared Errors (MSEs) value versus number of epochs for every training process.

Regression analysis between the bridge condition rating of the PCBB, RCBB and SBB bridges that were predicted by the ANN model and the corresponding bridge condition rating target is performed using the routine *postreg* using Matlab software. The format of this routine is $[m, b, r] = postreg\ (a, t)$, where m and b correspond to the slope and the intercept, respectively, of the best linear regression that relates the targets to the ANN outputs. If the fit is perfect, the ANN outputs are exactly equal to the bridge condition rating targets, and the slope is 1 and the intercept with the Y-axis is 0. The third variable, r, is the correlation coefficient

Table 4 Performance of network models for PCBB, RCBB and SBB bridges

Bridge superstructures	MSE	R^2-training	R^2-testing
PCBB	0.1004	0.6735	0.6188
RCBB	0.1125	0.9054	0.8637
SBB	0.1654	0.8177	0.7655

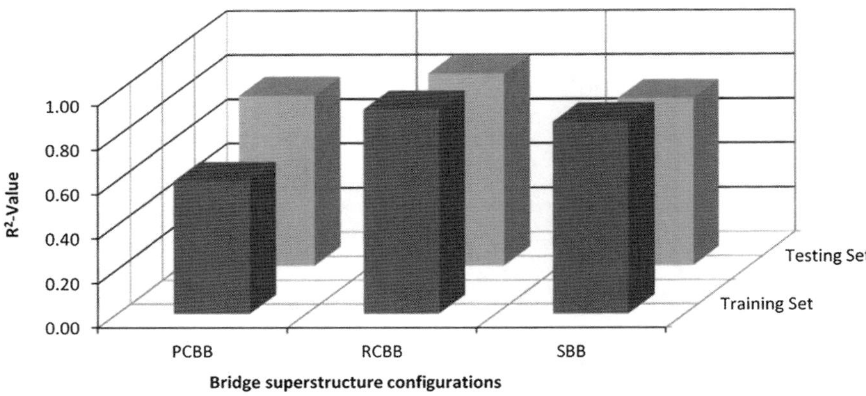

Fig. 2 Performance evaluation of ANN models based on R^2 value

between the ANN outputs and targets. It is a measure of how well the variation in the predicted bridge condition rating is explained by the target. If the r is equal to 1, then there is perfect correlation between the targets (t) and ANN outputs (a).

The training results show that an ANN trained with *trainlm* utilizing a single hidden layer with 4, 1 and 1 neurons in the hidden layer for PCBB, RCBB and SBB bridges, respectively, yields the best performance compared to other network structures, as indicated by the MSE and R^2 values. Table 4 shows the results of the best ANN models for condition rating modeling of PCBB, RCBB and SBB bridges.

The best ANN models are able to fit 67.35, 90.54 and 81.77 % of the impact of the primary bridge component condition on the entire condition rating of PCBB, RCBB and SBB bridges, respectively. For the testing set, ANN models were able to fit 61.88, 86.37 and 76.55 % for PCBB, RCBB and SBB, respectively. Figure 2 shows R^2 value of training and testing set data for modelling of PCBB, RCBB and SBB bridges based on its primary component condition ratings.

4 Conclusion

Four primary bridge components were used as independent variables or input parameters in modelling of PCBB, RCBB and SSB bridge condition ratings using MRA and ANN models. ANN models are able to fit 67.35, 90.54 and 81.77 % of the impact of the primary bridge component condition on the entire condition rating of PCBB, RCBB and SBB bridges. Meanwhile, the best MRA models able to fit 62.83, 91.77 and 86.18 % the proportion of the variance in the condition rating of PCBB, RCBB and SSB bridges are explained by all the primary bridge components in range of training data set. It can be concluded that, the primary component condition rating of RCBB and SSB have stronger influence to its entire bride condition rating. Meanwhile, the condition rating of primary components of PCBB shows slightly lower influence to the entire condition rating of the bridge compared to RCBB and SSB bridges.

The R^2 values of training set for the PCBB of ANN model was higher than the MRA models. Meanwhile, MRA models yield better result for RCBB and SBB bridges compared with ANN models. In terms of errors, the ANN models of PCBB, RCBB and SBB yield 0.1004, 0.1125 and 0.1654, respectively. Meanwhile, the best MRA models yield 0.1272, 0.1098 and 0.1331 for PCBB, RCBB and SBB bridges, respectively.

Referring to R^2 value and MSE of training set, the ANN model seems suitable to model PCBB bridge condition rating in respect to its primary bridge components. Meanwhile, for the condition rating of RCBB and SBB bridges, the multiple regression analysis seems more practical.

Acknowledgments The authors wish to acknowledge the financial support from the Ministry of Science, Technology and Innovation (MOSTI) of Malaysia for this work, which forms part of a project on the "Development of condition rating procedure of integral bridges" (Project Number: 06-01-02-SF0323). The authors also thank the Public Works Department of Malaysia for providing the data.

References

1. Othman MR, Choong CW, Heng LC (2002) Bridge management in Malaysia. Intertraffic Asia, Bangkok
2. Shuqing Z (2012) An uncertainty modeling for concrete bridge structures management. In: International conference on education technology and management engineering
3. Tserng HP et al (2009) Maintenance strategy for bridge components on the basis of performance. J Perform Constructed Facil 23:234
4. Yanev B (1998) The management of bridges in New York City. Eng Struct 20(11):1020–1026
5. Sasmal S et al (2006) Fuzzy logic based condition rating of existing reinforced concrete bridges. J Perform Constructed Facil 20(3):261–273
6. Chen MA (2008) Neural network approach for existing bridge evaluation based on grid. IEEE
7. Melhem H (1994) Fuzzy logic for bridge rating using an eigenvector of priority settings. In: IEEE conference proceedings of the first international joint conference of the North American fuzzy information processing society biannual conference; the industrial fuzzy control and intelligent systems conference; and the NASA joint technolo (NAFIPS/IFIS/NASA)
8. Wang YM, Elhag TMS (2008) Evidential reasoning approach for bridge condition assessment. Expert Syst Appl 34(1):689–699
9. Srinivasan R, Parlikad AK (2013) Value of condition monitoring in infrastructure maintenance. Comput Ind Eng 66(2):233–241

10. Saydam D, Bocchini P, Frangopol DM (2013) Time-dependent risk associated with deterioration of highway bridge networks. Eng Struct 54:221–233
11. Jain AK, Mao J, Mohiuddin K (1996) Artificial neural networks: A tutorial. Computer 29(3):31–44
12. Wang YM, Elhag T (2007) A comparison of neural network, evidential reasoning and multiple regression analysis in modelling bridge risks. Expert Syst Appl 32(2):336–348
13. Lee TL et al (2007) Neural network modeling for estimation of scour depth around bridge piers. J Hydrodyn Ser B 19(3):378–386
14. JKR (2003) Annual bridge inspection manual. Jabatan Kerja Raya, JKR Malaysia
15. Bolukbasi M (2004) Estimating the future condition of highway bridge components using national bridge inventory data. Pract Periodical Struct Des Constr 9:16
16. Adhikary BB, Mutsuyoshi H (2002) Use of fuzzy sets in condition rating of rc highway bridge structure. Proc Jpn Concr Instit 24(2):1525–1530
17. Pan N (2007) Forecasting bridge deck conditions using fuzzy regression analysis. J-Chin Inst Eng 30(4):593
18. Demuth, H.B. and M. Beale, *Neural Network Toolbox: for use with MATLAB*2000: Citeseer
19. Shamseldin A, Nasr A, O'Connor K (2002) Comparison of different forms of the multi-layer feed-forward neural network method used for river flow forecasting. Hydrol Earth Syst Sci 6(4):671–684

Evaluation of the True Behavior of the End Supports in the Carbajal de la Legua Old Bridge

A. González, E. Covián, M. Casero and M. Celemín

Abstract Bridges are prone to bearing deterioration due to aging, environmental causes and/or traffic overloading, and, as a result, a bridge originally designed as simply supported may be behaving differently due to changes in support conditions. The latter can lead to damage, not only to the bearings, but also to the rest of the structure due to unforeseen variations in the distribution of internal forces. In this paper, static load testing is used to evaluate the true behavior of the boundary conditions of an existing bridge and how they affect the overall structural response. The subject of investigation is a 16 m span concrete bridge subjected to the forces applied by a truck loaded with quarry aggregate. Static measurements are obtained for four positions of the truck and used to tune a mathematical model of the bridge until achieving an accurate resemblance with the experimental data. Boundary conditions are idealized with linear rotationary springs, which allow considering all possible degrees of rotational restraint at both end supports. It is shown how those values of the spring constants that best fit the measurements for a single loading case can be misleading and differ significantly from the optimal solution found combining all loading cases together.

A. González (✉)
School of Civil, Environmental and Structural Engineering,
University College Dublin, Belfield, Dublin 4, Ireland
e-mail: arturo.gonzalez@ucd.ie

E. Covián · M. Casero
Polytechnic School of Mieres, University of Oviedo,
Gonzalo Gutiérrez Quirós, s/n, 33600 Asturias, Spain

M. Celemín
Upper and Technical School of Agricultural Engineering, University of León,
Avda. de Portugal, 41, León 24009 León, Spain

A. Caner et al. (eds.), *Developments in International Bridge Engineering*,
Springer Tracts on Transportation and Traffic 9,
DOI 10.1007/978-3-319-19785-2_16

1 Introduction

Many structures are currently in need to assess their health and to detect possible damage caused by aging, environmental or loading factors (different or greater loads that the ones contemplated on the original design, fatigue, etc.). In the case of bridges, bearings are one of the most sensitive elements due to load concentration at that point and exposure to leakage and weather conditions, amongst other causes. Additionally, bearings are typically designed for a smaller return period than the bridge structure, which can cause early problems with respect to other bridge components, leading to their replacement or repair in an often costly intervention. The problems associated to bearings explain the increase in popularity of integral frame bridges (where bearings are omitted), over other structural forms. However, a considerable percentage of the bridge stock is supported on bearings and should receive periodic attention [1–5].

In this context, a static test is conducted in a structure located in Carbajal de la Legua, a small town near the city of León (Spain), in August 2013. Previously, in July 2013, a visit was made to the site to plan the tests and take preliminary measurements. Original designs of the construction were also consulted; although the available information was very limited due to the age of the bridge (built in 1933). The structure is a four-span straight bridge (Fig. 1) which is currently used only by pedestrians and, occasionally, by some cars and tractors. The reason for this situation is that the road has been re-routed through a new bridge that has been built just beside the one under investigation.

Static tests are performed on the third span from the left (Fig. 1) by means of a five-axle truck. Data from these tests is collected using surveying techniques (electronic tachymeter) and processed so that a mathematical model of the structure can be calibrated. A variety of models is tested allowing for boundary conditions to be modelled as pinned, fixed and linear rotationary springs. Varying the value of spring constants makes possible to consider a number of situations ranging from simply supported to fixed. Finally, the combination that best fits the measurements is considered to be representative of the structural response and bearing behavior.

Fig. 1 General view from the North side of the bridge

2 Description of the Site

In this section, the different elements of the test are described, beginning by basic features of the geometry of the structure. Then, surveying devices used are enumerated and, finally, the different loading conditions are listed.

2.1 Geometry of the Bridge

Figures 2 and 3 show the elevation and two cross-sections of the bridge respectively.

The bridge has a total length of approximately 64.5 m (16.42–16.22–15.96–15.95 m from East to West) and tests are focused on the third span ($L = 15.96$ m), where signs of deterioration can be visualized at the East support of the South beam (Fig. 4a). The scross-section (Fig. 3) can be described as an in situ double T section, composed of a horizontal top slab resting on two rectangular beams (similar for all spans). Both beams are placed with an eccentricity of 0.8 m with respect to the symmetry axis of the section, which leaves 0.95 m cantilevers at both sides. The width and depth of the top slab are 3.8 and 0.39 m respectively, constant along the structure. Beam height varies from 1.1 m at the supports (Section A-A′ in Fig. 3) to 1 m (Section B-B′ in Fig. 3) after a distance of approximately 1.09 m

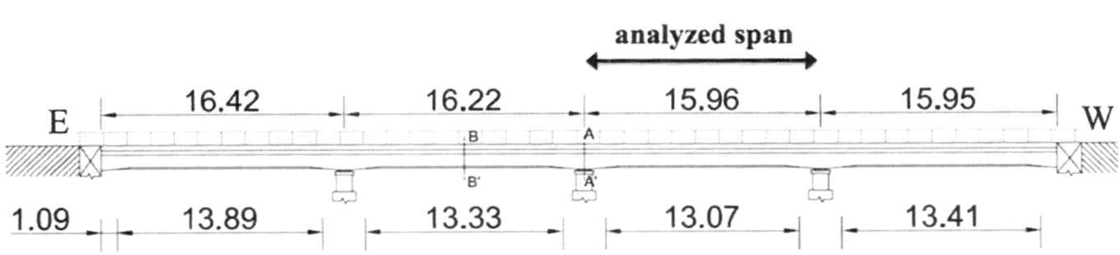

Fig. 2 Elevation (all dimensions in m)

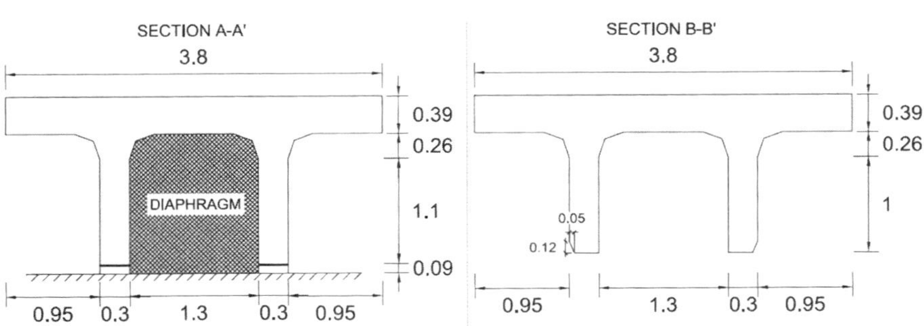

Fig. 3 Cross section (all dimensions in m)

Fig. 4 Supports: **a** signs of deterioration at the support, **b** detail of the bearing

from the edge of the bearing (Fig. 2), remaining with a constant height in the rest of the span. At that transition point, a bevel on the exterior face of the beams appears and it continues over their remaining length with dimensions of 5 cm (horizontal) and 12 cm (vertical) (Section B-B′ in Fig. 3). The connection of the top slab to the beams is made over a depth of approximately 0.26 m. Therefore, the total depth of the deck ranges from 1.65 to 1.75 m at the supports, which corresponds to a depth-span ratio of 0.103 and 0.109 respectively (depth $\cong L/10$ in both cases).

At the supports, a trapezoidal concrete platform serves to level the bearings, which consist of a metal layer. A photo of the bearing detail is provided in Fig. 4b. Dimensions of the leveling platforms are 24 cm for the top base, 47 cm for the bottom base and 8.5 cm for its height. As for the bearing, it is 24 cm long and 1.3 cm high; on the transversal direction both elements have the same width as the beams (30 cm). Initially, it is assumed that spans are independent since four bearings can be found at all piers (two adjacent spans and two beams per span). Nevertheless, presence of dirt and soil at the supports encourage taking these assumptions and the afore-mentioned dimensions with caution. Finally, concrete diaphragms between beams can be noticed at all supports.

2.2 Measurement Devices

First of all, the overall dimensions of the structure are taken using a laser distance meter and a tape measure. The distance meter is a model Leica DISTO D8 with 0.5 mm uncertainty [6] while the tape measure has an accuracy of 1 mm. These instruments are also used to take the measurements needed to place the surveying targets on the structure.

Surveying measurements are taken using an electronic tachymeter (total station) Leica TCRP1201 in conjunction with sticker black and white targets. Specifications for this device are the following: 1 mm + 1.5 ppm in distance and 1″ in horizontal and vertical angles [7]. Targets are placed on the exterior face of the beams and their location is symmetric with respect to mid-span. They are located in a quasi-horizontal line that is marked using the print left by the formwork. Targets are distanced 0.5 m in the central part of the beams and 1 m near the supports.

2.3 Loading Conditions

The vehicle used is a five-axle truck (Fig. 5), with two wheels per axle, except for the second one that has four wheels. Distances between wheels (center to center) for each axle are as follows: 2122 mm (1st axle), 350–1711–350 mm (2nd axle), 2040 mm (3rd axle), 2040 mm (4th axle), 2040 mm (5th axle); and distances between axles are: 3.6 m (1st to 2nd), 2.7 m (2nd to 3rd axle), 1.31 m (3rd to 4th axle) and 1.31 m (4th to 5th axle). The footprint of the tires on the deck surface measures approximately 25 cm by 16–22 cm. The truck is loaded using aggregate from a nearby quarry, adding up to a total gross vehicle weight of 40.29 t. The individual axle weights of the 1st, 2nd, 3rd, 4th and 5th axles are 7.26, 9.06, 8.02, 8.17 and 7.78 t respectively.

Table 1 summarizes the four testing scenarios defined by the position of the second axle of the truck along the span and the orientation of the truck. In all cases, the truck is applied centered with respect to the symmetry axis of the bridge

Fig. 5 Testing vehicle

Table 1 Static loading cases

Scenario	Truck orientation (5th axle–1st axle)	Truck 2nd axle position	Min. distance between wheel footprint and edges of top slab (Fig. 3) (North/South in cm)[a]
I	West-East (right Fig. 5)	Mid-span	60/79
II	West-East	L/4 to East support	72/69
III	East-West	Mid-span	75/67
IV	East-West	L/4 to West support	72/69

[a]Distance measured from the exterior face of the wheel to the edge of the deck

cross-section (i.e., it can be assumed that the truck load will be equally distributed to the left and right of the symmetry axis in Fig. 3).

3 Field Measurements

In this section, steps followed during field work are first described and then, measured data is presented. Before beginning with surveying measurements, it is necessary to choose the points where the tachymeter will be stationed and calculate their coordinates. For this purpose, a surveying method known as "traverse" [8] is implemented. After selecting two station points, one at each side of the bridge, a third one is chosen in order to close the itinerary. Then, random coordinates are assigned to one of these points and, starting from that same point, observations of angles (horizontal and vertical) as well as distances are consecutively taken from each point to the other two. Finally, a new set of measurements is taken from the starting point. Using all these data, unknown coordinates for the two station points are calculated.

Measurements on the bridge are taken using another surveying method known as "radiation" that consists on placing the device on a point of known coordinates and measuring angles (with respect to a fixed origin) and distances to the target points. Coordinates for those points can be determined using the acquired data. Measurement sequence is the following: (i) unloaded bridge measured from station point located North of the bridge, (ii) loaded bridge measured from North station point, (iii) loaded bridge measured from station point located South of the bridge and (iv) unloaded bridge measured from South station point. Between steps (i) and (ii) truck is put into position and between (iii) and (iv) it leaves the bridge. For each target, coordinates are calculated three times and then averaged in order to minimize accidental errors inherent to the measurement process. Finally, displacements are calculated using surveying techniques for static scenarios (I–IV).

The deflections measured for the scenarios in Table 1 are presented in Fig. 6.

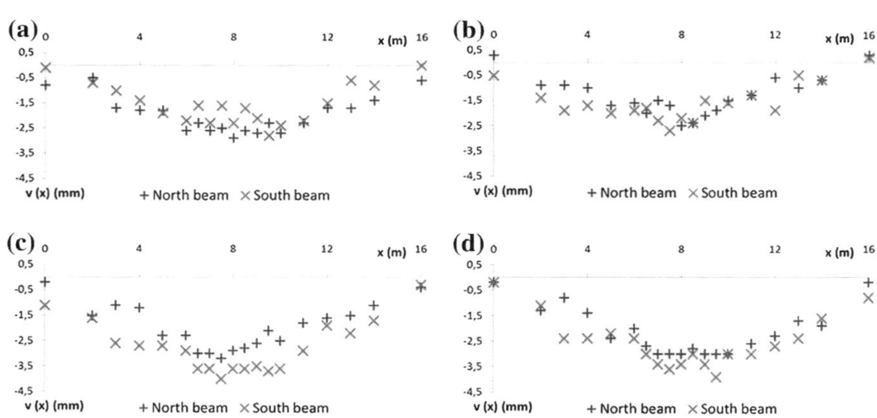

Fig. 6 Measurement profiles: **a** I, **b** II, **c** III and **d** IV

4 Model Calibration and Testing

In the analysis of the measured deflections, one-dimensional mathematical models are used to represent the response of each half of the bridge section in Fig. 3, which is basically a T-beam. It is assumed that the truck load is equally shared by the two T-beams composing the section. Boundary conditions at the supports and flexural stiffness *EI* (modulus of elasticity multiplied by inertia) of each beam are the unknowns in these models. For simplification purposes, *EI* is assumed to be constant all along the beam. Values for the parameters of the models are estimated by minimizing the objective function which can be described as the root mean square error (RMSE) between measured displacements (Fig. 6) and those predicted by the model. Regarding flexural stiffness, it is necessary to say that based on a gross inertia of 0.228 m^4 (estimated from section B-B′ in Fig. 3) and a typical modulus of elasticity of reinforced concrete of 30 GPa, an *EI* value of 6840 MN m^2 can be derived for each beam. However, the original strength of the concrete and the amount of reinforcement employed in the section is unknown, the loss of properties due to age deterioration is uncertain and the inertia is not constant along the beam. Therefore, the solution is expected to lie somewhere between 4560 and 9120 MN m^2.

Typical pinned-pinned, pinned-fixed, fixed-pinned and fixed-fixed support conditions are tested first. Here, the sole unknown parameter of the model is *EI* given that boundary conditions are imposed. In the labeling of these boundary conditions (pinned-fixed, fixed-pinned, etc.), the first and second words refer to West and East supports, respectively. Results for both beams and scenarios I to IV are presented in Table 2.

Table 2 Estimation of stiffness for beam model with pinned or fixed support conditions

Stiffness (MN m^2)	North beam				South beam			
	Pin-pin	Pin-fix	Fix-pin	Fix-fix	Pin-pin	Pin-fix	Fix-pin	Fix-fix
Scenario I	4960	2194	1811	959	6077	2689	2209	1169
Scenario II	6830	2642	3022	1430	5922	2324	2611	1255
Scenario III	4879	1770	2165	939	3579	1302	1592	694
Scenario IV	4400	1929	1729	929	3770	1669	1474	801
Mean	**5267**	**2134**	**2182**	**1064**	**4837**	**1996**	**1972**	**980**
Standard dev.	*1071*	*381*	*591*	*244*	*1346*	*626*	*534*	*274*
COV	**0.20**	**0.18**	**0.27**	**0.23**	**0.28**	**0.31**	**0.27**	**0.28**
RMSE (mm)	**1.32**	**1.59**	**1.95**	**1.53**	**1.51**	**2.10**	**2.11**	**1.99**

Bold is chosen for mean values (which correspond to the final estimate of the parameters), COV and RMSE(the two parameters used to adopt a final solution), which are the ones widely used for the results discussion

Italic is chosen for standard deviation values which, although corresponding to the overall analysis of allloading scenarios at once, are not heavily mentioned in the results discussion

Table 2 shows *EI* estimates for each loading case and mathematical model. Mean, standard deviation and coefficient of variation (COV = standard deviation/mean) are provided to evaluate the consistency of results throughout the four tests. The last row of Table 2 corresponds to the root mean square error in mm, added for all four scenarios. It can be noticed that for both North and South beams, the more realistic *EI* values are obtained for the case of the simply supported beam. For the rest of the support conditions, the resulting *EI* values lie significantly below the range deemed to be reasonable (4560–9210). Similarly, minimum RMSE corresponds to the pinned-pinned so it is concluded that a simply supported beam is closer to the measurements than any other model in Table 2. However, COV is still significant (i.e., relative large variation in *EI* depending on the loading case under consideration) and it is decided to introduce rotationary springs at the supports in an attempt to further improve both COV and RMSE.

As a result, the beam model has now three unknowns: *EI*, West spring constant (K_{West}) and East spring constant (K_{East}). Thus, the goal is to find the combination of values of these three parameters that provides the minimum RMSE. There is a wide range of values of spring constants, therefore, those values in range with power 0 are deemed to basically correspond to the pinned-pinned case (no significant differences in the displacement response) and those values in excess of power to the 6 agree with the fixed-fixed case (for higher values of the spring constant, differences in the profile with the fixed-fixed case are negligible). As a consequence, spring constants are considered to vary from 10^0 to 10^6 MN m/rad. Optimal sets of values that minimize RMSE are obtained for the parameters and each load scenario. Figure 7 shows two contour plots corresponding to the North and South beams, considering one of the parameters constant and equal to the value that provides the minimum RMSE.

It is difficult to appreciate a minimum in Fig. 7, given that is located in a flat valley region for both beams. Nevertheless, combination of parameters that gives the minimum value of RMSE is marked with a single black dot. Figure 7a corresponds to the North beam, for load scenario I, with K_{East} constant and equal to its optimum value. Similarly, Fig. 7b represents the variation of the objective function for the South beam and same conditions as Fig. 7a. Table 3 summarizes the

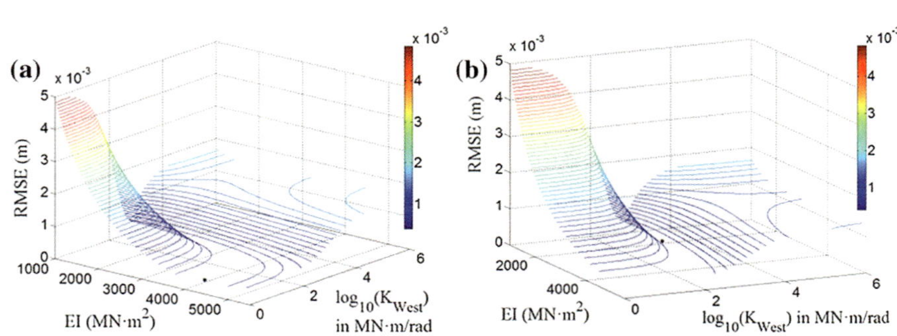

Fig. 7 RMSE versus K_{West} and *EI*: **a** North beam, **b** South beam

Table 3 Estimation of section stiffness and spring constants at the supports

	North beam			South beam		
	Stiffness (MN m^2)	K_{West} (MN m/rad)	K_{East} (MN m/rad)	Stiffness (MN m^2)	K_{West} (MN m/rad)	K_{East} (MN m/rad)
Scenario I	4310	$10^{0.25}$	$10^{2.375}$	2110	$10^{2.8125}$	$10^{2.5}$
Scenario II	2630	$10^{2.6875}$	$10^{2.75}$	3360	$10^{3.3125}$	10^{0}
Scenario III	2020	$10^{2.3125}$	$10^{2.75}$	3470	10^{0}	$10^{1.5}$
Scenario IV	2560	10^{0}	$10^{3.125}$	3000	10^{0}	$10^{2.5}$
Mean	**2880**	**174**	**674**	**2985**	**676**	**166**
Standard dev.	*992*	*230*	*466*	*617*	*968*	*174*
COV	***0.34***	***1.32***	***0.69***	***0.21***	***1.43***	***1.04***
RMSE (mm)	**1.22**			**1.45**		

Bold is chosen for mean values (which correspond to the final estimate of the parameters), COV and RMSE(the two parameters used to adopt a final solution), which are the ones widely used for the results discussion

Italic is chosen for standard deviation values which, although corresponding to the overall analysis of allloading scenarios at once, are not heavily mentioned in the results discussion

combinations of parameters (EI, K_{West} and K_{East}) that provide minimum RMSE values for each load case. The value of RMSE in Table 3 corresponds to the root mean square error in mm, added for all four scenarios.

Given the variability of the results and the unrealistically low values of EI shown in Table 3, in order to find a more reliable solution, it was decided to take an "L-curve" approach, which would allow minimizing not only the RMSE but also the COV of stiffness predictions for all load scenarios. This approach consists of testing a number of combinations of values for K_{East} and K_{West}; while computing the EI value that provides the minimum RMSE for each load scenario. Thus, for each pair of spring constants, four EI values would be obtained (one per loading case), which makes possible to determine their mean, standard deviation and COV. Figure 8 shows the COV in the x-axis and the total RMSE in the y-axis, with each point of the graphic corresponding to a specific combination of values of spring constants. In order to decide which point represents the best combination of parameters for the beam models, Pareto criterion is applied. This criterion states that a solution (y_i) dominates another (y_j) if all vector components of y_i are not greater than the corresponding components of y_j (1st condition) and at least one component of y_i is strictly smaller than the corresponding component of y_j (2nd condition). For the case under investigation, this means that a solution A (COV_A, $RMSE_A$) would dominate another solution B (COV_B, $RMSE_B$), if:

- $COV_A \leq COV_B$ & $RMSE_A \leq RMSE_B$ (1st condition); and
- (i) $COV_A < COV_B$ or (ii) $RMSE_A < RMSE_B$ or (i) and (ii) at the same time (2nd condition).

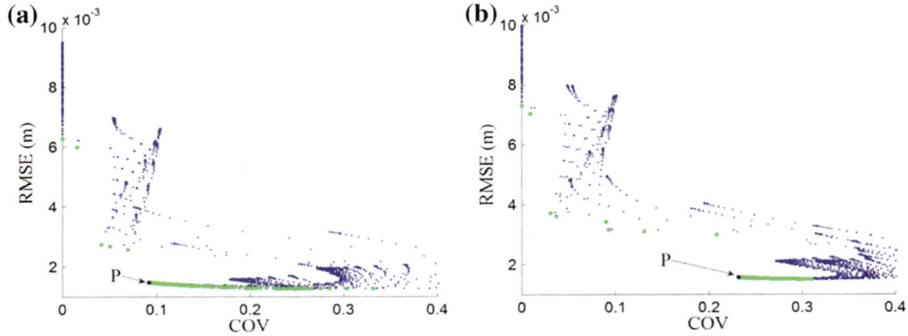

Fig. 8 RMSE versus COV: **a** North beam, **b** South beam

All solutions that are not dominated by any other are said to be the optimal ones. Points that fulfill Pareto criterion are marked with full green dots in Fig. 8.

Amongst all optimal solutions shown in Fig. 8, the ones chosen as final values for each beam model are those closer to the origin point, which would represent the perfect prediction (no variation coefficient and null RMSE). Due to noise in the measurements, selection of optimum point is not clear; even when applying Pareto criterion it could be possible that the points selected as dominant may be affected by noise. If points closer to the origin (avoiding isolated points) are selected, values of 10^0 MN m/rad for K_{West}, $10^{0.5}$ MN m/rad for K_{East} and 4927.5 MN m^2 for EI are obtained in the North beam (point P marked in Fig. 8a); and values of $10^{0.375}$ MN m/rad for K_{West}, $10^{0.25}$ MN m/rad for K_{East} and 4582.5 MN m^2 for EI are obtained for the South beam (point P marked in Fig. 8b). EI for both beams falls within the lowest values of the theoretical range (4560–9210), which can be expected for such an old bridge, cast in situ and given the construction practice, materials and quality controls at the time. If results in Fig. 8 are compared to values in Table 2 corresponding to the simply supported case, the trade-off between COV and RMSE is clear for both beams. For the North beam, COV decreases (from 0.20 in Table 2 to 0.09 in Fig. 8a) while RMSE increases (from 1.32 to 1.5 mm). Similarly, for the South beam, COV decreases from 0.28 in Table 2 to 0.23 in Fig. 8b and RMSE increases slightly (from 1.51 to 1.6 mm). It should be noticed that values in Table 2 have been calculated similarly to those in the "L-curve" approach of Fig. 8, since the boundary conditions used are just particular cases of the spring constants. As for Table 3, the sets of COV and RMSE values are not directly comparable to those derived from Fig. 8. While values in Table 3 correspond to minimizing each loading scenario individually, i.e., leading to different spring constants for each scenario, the optimal result of the "L-curve" approach in Fig. 8 seeks to minimize all scenarios at once, i.e., each point represents a unique set of spring constants being tested for all scenarios.

The values of EI, K_{East} and K_{West} are relatively homogeneous for both beams, being the boundary conditions very similar to a simply supported beam. It is true that *EI* of the South beam is slightly smaller than that of the North beam, which appears to be in agreement with the preliminary visual inspection (i.e., damaged area in Fig. 4a); however, the impact on the structural response is not significant. Displacements are small thanks to the large section of the bridge and the low values of spring constants resemble the response of simply supported beams in both North and South models.

5 Conclusions

The health of an old bridge located near León (Spain), in particular its stiffness and support conditions, have been assessed via surveying measurement of deflections due to static loading tests. The paper has shown that it is important to take into account the uncertainty associated to the measures when trying to resemble them with mathematical models of the bridge response. The use of only one static loading test can be misleading due to inaccuracies in measurements and models, and the reliability of the results can be improved by combining a number of loading cases.

Acknowledgments The authors wish to express their gratitude for the support received, under Grant No. BIA2011-26915, from the Spanish "Programa Nacional de Proyectos de Investigación Fundamental, del VI Plan de Investigación Nacional Científica, Desarrollo e Innovación Tecnológica 2008–2011" towards this investigation. They would also like to thank the support offered by the "Diputación Provincial de León" and, in particular, to Emilio Barnechea, for facilitating the access to the structure and authorizing its closure during the tests; and by POSADA Organización, S.A. for supplying the truck and quarry aggregate. Finally, a special thanks to Miryam Valle Feijóo for her assistance in the field work.

References

1. Sianipar P, Adams T (1997) Fault-tree model of bridge element deterioration due to interaction. J Infrastruct Syst 3(3):103–110
2. Farhey D, Naghavi R, Levi A, Thakur A, Pickett M, Nims D, Aktan A (2000) Deterioration assessment and rehabilitation design of existing steel bridge. J Bridge Eng 5(1):39–48
3. Attoh-Okine NO, Bowers S (2006) A Bayesian belief network model of bridge deterioration. Proc ICE-Bridge Eng 159(2):69–76
4. Frangopol D, Lin K, Estes A (1997) Life-cycle cost design of deteriorating structures. J Struct Eng 123(10):1390–1401
5. Kawashima K, Unjoh S (1992) The damage of highway bridges in the 1995 Hyogo-Ken Nanbu Earthquake and its impact on Japanese seismic design. J Earthq Eng 1(3):505–541
6. Leica Geosystems AG (2011) DISTO ® 8 user manual and data sheet. Leica Geosystems AG, Heerbrugg (Switzerland)
7. Leica Geosystems AG (2004) LEICA TPS1200 series. User manual and data sheet. Leica Geosystems AG, Heerbrugg (Switzerland)
8. Ghilani CD, Wolf PR (2006) Adjustment computations spatial data analysis. Wiley, New Jersey

Bridge Instrumentation: Needs, Options, and Consequences

A. Shamsabadi, F. Abazarsa, S.F. Ghahari, R. Nigbor and E. Taciroglu

Abstract Seismic design procedures are in transition from strength-based to performance-based approaches. Faster and larger computers and improved nonlinear behavior models have paved the way. Nevertheless, validations of these predictive models have typically been confined to experiments conducted at the component scale. Controlled experiments at the system level are scarce. At the present time, the next best option for model validation is to use dynamic data collected from instrumented structures. These data include recorded accelerations due to impact excitations, ambient forces, weak or distant earthquakes, and strong ground motions. Different techniques are needed to extract useful information for such diverse types of data. Herein, we examine a long-span bridge located in California-which is instrumented through the joint efforts of Caltrans and the California Strong Motion Instrumentation Program (CSMIP)-as a case study for demonstrating how and what types of information may be extracted from recorded

Title here pays homage to work by D.M. Boore and J.J. Bommer, titled "Processing of strong-motion accelerograms: needs, options and consequences," and published in Soil Dynamics and Earthquake Engineering, (2005, vol. 25, pp. 93–115).

A. Shamsabadi
Office of Earthquake Engineering, California Department of Transportation, Sacramento, CA
e-mail: Anoosh_shamsabadi@dot.ca.gov

F. Abazarsa · S.F. Ghahari · R. Nigbor · E. Taciroglu (✉)
University of California, Los Angeles, CA 90095, USA
e-mail: etacir@ucla.edu

F. Abazarsa
e-mail: abazarsa@gmail.com

S.F. Ghahari
e-mail: ghahari@gmail.com

R. Nigbor
e-mail: nigbor@ucla.edu

A. Caner et al. (eds.), *Developments in International Bridge Engineering*,
Springer Tracts on Transportation and Traffic 9,
DOI 10.1007/978-3-319-19785-2_17

data that bear consequences for engineering practice. Some of the highlights of this study include the observation of some inconsistencies between data and metadata, calculation of wave delays, and response prediction with updated finite element models.

1 Introduction

It is fair to state that the bridge engineering community had already reached a high level of competency in developing accurate simulation models based on decades of research and field observations on superstructure components as well as foundation elements (e.g., abutments, pile foundations, etc.). However, there still remain uncertainties that arguably introduce significant errors into responses predicted for strong seismic events. At the present time, soil nonlinearities and soil-structure interaction (SSI) effects are the primary sources of these modeling uncertainties. While there have been numerous past efforts to quantify the SSI effects for bridge support structures through component testing (for example, of large-scale specimen abutments [1] and piles [2]), their true behavior is coupled with the superstructure. Therefore, direct or indirect comparisons with measurements from real-life structures during strong ground shaking present the best opportunity to validate any predictive model that aims to mimic their responses.

In this study, we present results of finite element modeling and model–updating studies for an instrumented long-span bridge in California. The primary objective has been to develop quantitative comparisons of dynamic responses predicted through the use of high-fidelity state-of-the-art bridge models and real–life data obtained through the Caltrans/CSMIP Bridge Instrumentation Program [3, 4]. These comparisons are intended to determine what the *needs* (regarding instrumentation and modal identification), *options* (regarding model updating), and *consequences* (regarding what useful information can be gleaned from recorded data).

In what follows, we first introduce the specimen bridge and its instrumentation. We then present a chronicle of our studies with data collected at this bridge during several strong earthquakes and an ambient vibration survey, which include pre-processing of the strong motion data, modal identification and finite element modal updating, and finally a blind response prediction using the updated model. For brevity, only a summary of the said studies is presented here, and further details can be found in [5].

2 Description of the Investigated Bridge

The bridge selected for detailed investigation is the *Eureka–Samoa Channel Bridge* (henceforth referred to as the *Samoa Channel Bridge*, or simply SCB). The SCB was chosen specifically because it features a "free-field" station, another nearby geotechnical downhole array, an instrumented pile foundation, and seven

Fig. 1 The Samoa channel bridge

significant earthquake events recorded to date—the most important event being the 2010 Ferndale Earthquake ($M_w = 6.5$), which induced a Peak Structural Acceleration (PSA) of 0.37 g on the bridge deck.

The SCB carries Route 225, linking the city of Eureka to Samoa Peninsula (Fig. 1). It was constructed in 1971 (construction started in 1968) and underwent a seismic safety retrofit in 2006. The bridge is approximately 2506 ft long and 34 ft wide. Its superstructure comprises 6.5 in-thick concrete deck slabs resting on four pre-stressed precast concrete I-girders with intermediate diaphragms. The composite deck is supported on concrete bent-cap and hexagonal single-columns and seat-type abutments. The bridge consists of 20 spans. The typical span length is 120 ft except the main channel, which is 225 ft-long, and extends from the centerline of pier 8 to the centerline of pier 9. The 150 ft-long concrete I-girders of the superstructure begin at pier 7 and pier 10, and are cantilevered 30 ft past piers 8 and 9 into the main-channel crossing span. The 165 ft-long pre-stressed precast concrete I girders resting atop the cantilevered portions cross over the main-channel.

The layout of instrumentation is shown in Fig. 2. The SCB was instrumented first in 1996 through installing 24 accelerometers on the bridge and 3 accelerometers at a nearby free-field site. During structural retrofit at 2006, in an unusual procedure, accelerometers were added at depths of 34 and 54 m in one of the cast-in-place piles that were installed at Pier 8. At the time there was no way to accurately establish the orientation of the horizontal in-pile sensors (channel # 28, 30, 31, and 33), so their orientations have not been precisely known.[1] Due to the complexity of its geometry and site soil conditions, this bridge has not been studied with requisite detail until 2012, when Caltrans commissioned a comparative study of finite element model predictions and its measured accelerations obtained from the CSMIP seismic instrumentation [5].

Based on Caltrans classifications, the Samoa Bridge nominally belongs to the group of "Ordinary Standard Bridge" structures, provided that soil conditions are

[1]Personal communication with T. Shakal (California Geological Survey) and P. Hipley (Caltrans).

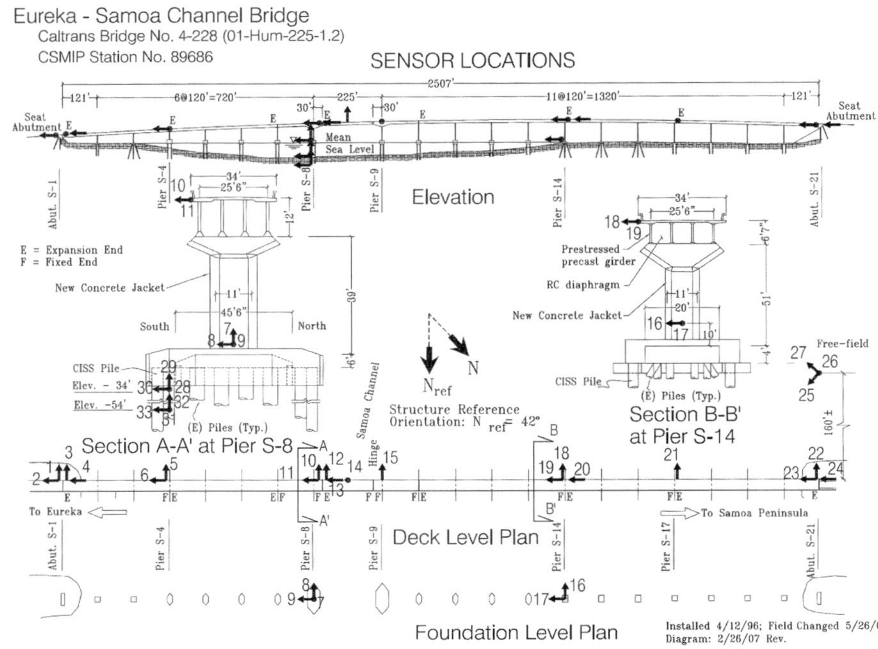

Fig. 2 Instrumentation layout of Samoa channel bridge (CSMIP)

normal, in which case the underlying soil can be assumed rigid and SSI can be neglected. However, in §6.1.1 of the Caltrans Seismic Design Criteria [6] document, it is stated that "[w]hen bridges are founded on either stiff pile foundations, or pile shafts and extend through soft soil, the response spectrum at the ground surface may not reflect the motion of the pile cap or shaft. In these instances, special analysis that considers soil pile/shaft kinematic interaction is required and will be addressed by the geo professional on a project specific basis." In addition to kinematic interaction, soil-flexibility must be addressed if the soil is deemed "poor" or "marginal." Since this caveat is true for SCB (it has pile foundations), SSI effects should be considered. Nevertheless, it is not specifically stated how these effects should be quantified and modeled, for example, in the seismic safety assessment of this bridge. As it will be shown in the following sections, (i) both inertial and kinematic SSI effects play significant roles in SCBs response to seismic events, (ii) these effects can be quantified using data recorded during strong motion event, (iii) and finally, they can be modeled in a computationally efficient manner (i.e., without having to resort to, for example, three-dimensional finite element modeling of the soil media surrounding/supporting the bridge).

To this end, first, a detailed Finite Element model of SCB was created that featured state-of-the-art soil-pile springs [7, 8] based on available structural drawings and geotechnical data. Not surprisingly, responses predicted with this initial model at the sensor locations did not match the response signals recorded during various earthquakes [5] (in fact, a true comparison is not even possible, because the foundation input motions are not directly recorded due to soil-structure interaction effects), which necessitated the calibration (updating) of the said initial FE model. Various stages of this procedure are described in the following sections.

3 Needs: Instrumentation and Modal Identification

Data recorded at instrumented bridges provide valuable information, provided that the associated metadata are reliable. As an example, the displacement time-histories of the data recorded during the Ferndale earthquake, 2007, on pier 8 of SCB are shown in Fig. 3. Also, the Free-Field Motions (FFMs) that are recorded at channels 25 and 27, and rotated in the longitudinal and transverse directions of SCB are shown here. As seen, the patterns of the signals recorded by the in-pile sensors are different from others.

As it was discussed earlier, the true orientations of these in-pile channels are not known, and thus they cannot be used unless corrections are made. However, thanks to inertial soil-pile-structure interaction, these sensors do contain the dynamic characteristics of the entire bridge-foundation system. Because SCB is a straight bridge, its modal characteristics in two transverse and longitudinal directions (principal directions) should be independent. Therefore, if the in-pile sensors are actually oriented along the directions shown in Fig. 2, then they must reflect only transverse and longitudinal natural frequencies of the bridge-foundation system separately in each channel. However, as it may be seen in Fig. 4 (left), both channels exhibit the same dominant frequency, which is the first natural frequency of the bridge in the transverse direction. Hence, we can identify true orientation by rotating these two channels until they display distinct dominant frequencies. In Fig. 4 (right), it is observed that, by rotating almost 30° (clockwise), we indeed

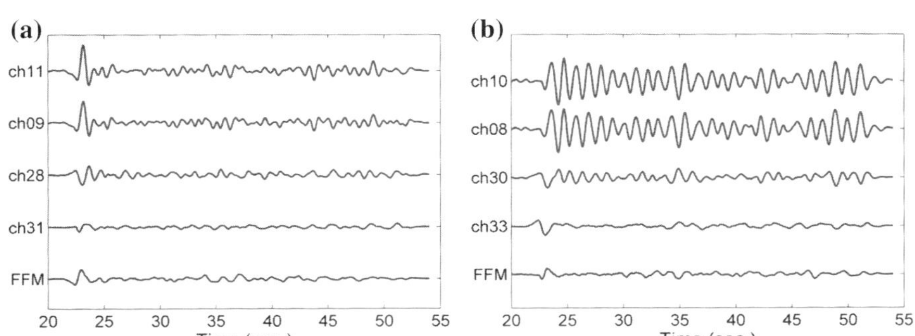

Fig. 3 Signals recorded during Ferndale earthquake (2007) at the free-field and pier no. 8 of SCB: **a** longitudinal and **b** transverse directions

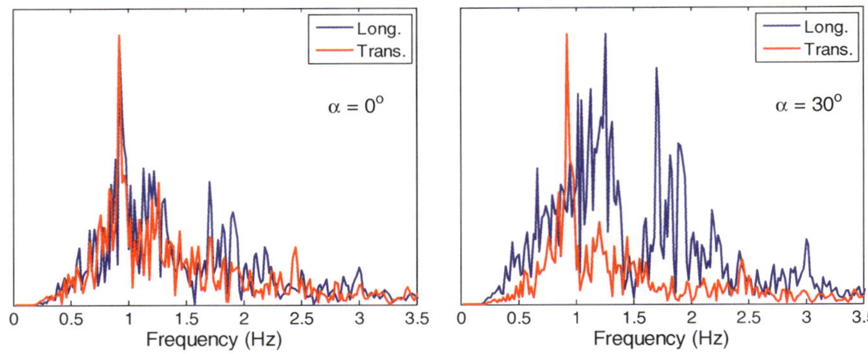

Fig. 4 Fourier spectra of channels 28 and 30 at two different orientations

obtain such a frequency separation. Data from other earthquakes suggest the same orientation (omitted here for brevity). For the signals recorded by the pile-tip sensors, this process is not helpful, because these channels are mostly affected by input motions rather than by vibrations of the superstructure. However, CGS offered corrections to their orientations based on the similarity to the recorded FFMs.[2] The approach described above to detect the true orientations of signals is somewhat qualitative. A veritable and more precise method based on cross-correlation analysis is described and validated in [5] and details are omitted here for brevity.

This initial analysis of data and metadata suggests the following: (i) meta-data supplied along with records may also contain uncertainties, and the analyst would do well to inspect them in detail, and seek corrections as needed; (ii) motions recorded at the foundations (even at the extreme ends of the piles) may exhibit frequency content from the entire bridge system, which means that there are inertial SSI effects present, and the recorded motions *cannot be treated as input motions*. This latter item also implies that input-output modal identification methods are not applicable here and moreover, conventional output-only identification methods (e.g., the frequency domain decomposition method, the natural excitation technique, the eigen-realization algorithm, etc.) are *inappropriate*. This is because these methods/algorithms are based on the assumption that the unknown input is stationary whitenoise, yet the recorded signals on SCB were generated by earthquake sources, which are non-stationary—especially during their strongest time-intervals. Nevertheless, there are recently developed output-only identification methods that remove the stationarity restriction, so they can be applied to identification problems that feature SSI effects [9, 10].

Using the corrected signals, and an appropriate output-only technique, the modal characteristics of SCB were identified first using signals recorded during earthquakes. While the said identification approach yielded natural frequencies,

[2]Personal communication with M. Huang, California Geological Survey.

Table 1 Identified natural frequencies (Hz) from different events

Direction	Mode no.	Ambient survey 2014	Willow Creek 2008 (PSA = 0.032 g)	Ferndale 2010 (PSA = 0.370 g)
Transverse	1	1.00	0.91	0.68
	2	1.16	1.10	0.77
	3	1.33	1.14	0.86
	4	1.48	1.30	0.98
	5	1.60	1.47	1.02
	6	1.74	1.55	1.13

the mode shapes could not be identified reliable for this multi-span bridge, which is essentially only instrumented at its main channel-crossing span (Fig. 2). This necessitated an ambient vibration survey of the bridge using a dense sensor array, which was carried out in collaboration with a team from the University of British Columbia, on June 6, 2013 [11]. This ambient vibration survey, and the ensuing modal identification studies revealed several important findings: (i) bridges exhibit amplitude-dependent modal characteristics, and their natural frequencies typically decrease with increasing shaking intensity; (ii) the said variation is usually recovered once the excitation subsides, except for strong/damaging earthquakes (this will be demonstrated later for SCB); (iii) the sensor density in presently instrumented bridges are generally inadequate for reliable identification of mode shapes, or even many of the important higher modes entirely.

The first point made above here is clearly displayed in Table 1, which displays the natural frequency identified from ambient test data, as well as a medium and a strong earthquake.[3]

The second point is illustrated in Fig. 5, which displays the sudden drop in the first transverse natural frequency of SCB around 40 s into the 2010 Ferndale earthquake. As this figure shows, at the first 20 s, system behaves similar to ambient excitations. When the severe waves arrive, the natural frequency drops from 1.45 to 1.29 Hz (i.e., the frequency the system exhibits under weak earthquakes). Its frequency then continues to decrease to 0.89 with the arrival of the second wavefront. After this, the bridge reaches its lowest frequency for almost 10 s, and subsequently its frequency is starting to recover the large vibrations diminish.

The last point made above is illustrated below for SCB, where Fig. 6 displays the mode shapes identified using a dense sensor array as well as those identified using only the existing permanent CSMIP sensors. As it may be seen from this figure, while the accuracy of modal amplitudes at the sampling locations are not necessarily inaccurate, the spatial resolutions of the identified mode shapes are far from being adequate. This statement can be quantified through the use of Modal Assurance Criterion ($0 < \text{MAC} < 1$) [5], which measures the similarity between two vectors (mode shapes). If two mode shapes are identical, then their MAC

[3]Results from several other weak earthquakes (besides Willow Creek 2008) are available in [5], and yield similar results.

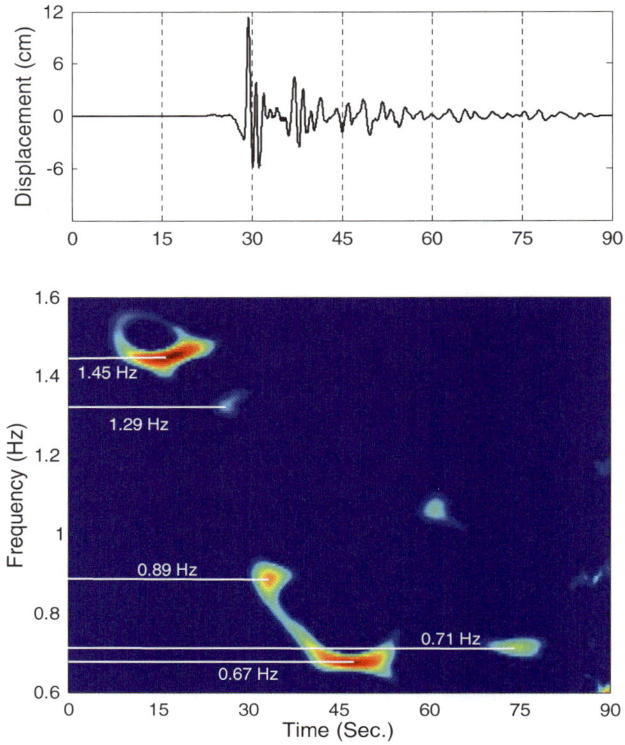

Fig. 5 Displacement time-history (*top*) and time-frequency representation of the response recorded at channel #6 during the 2010 Ferndale earthquake

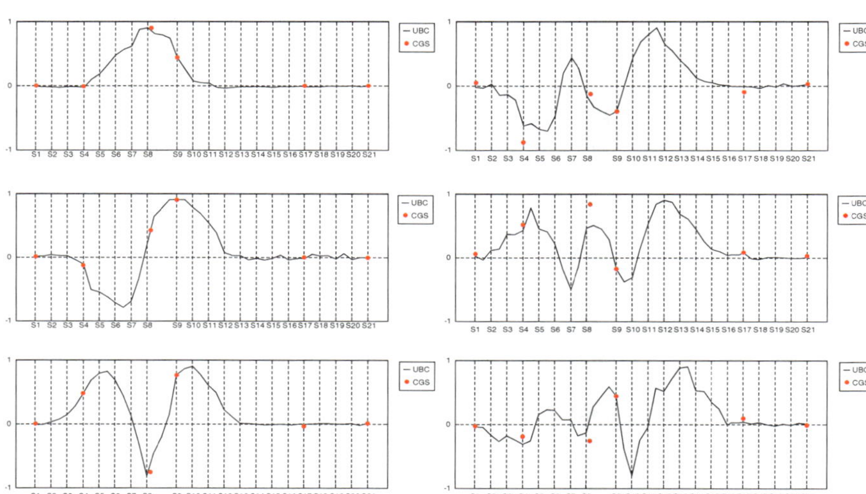

Fig. 6 Comparison of mode shapes identified from a dense sensor array (*black continuous lines*) and from the existing CSMIP sensor array (*red dots*)

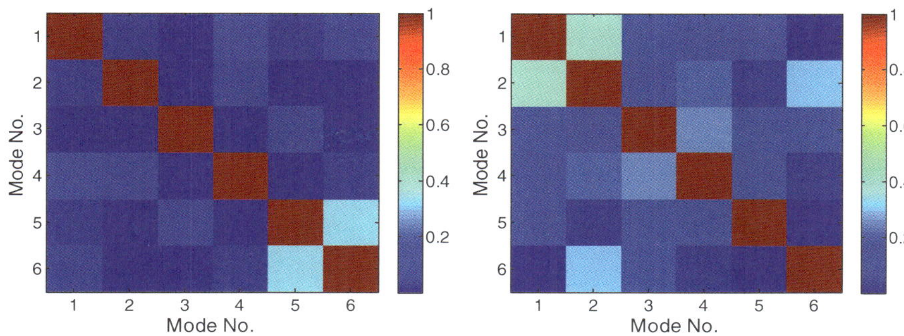

Fig. 7 MAC indices among the transverse mode shapes of SCB identified from a dense sensor array (*left*) and the existing CSMIP sensor array (*right*)

index is equal to 1. Figure 7 (left) displays the MAC indices among modes identified using a dense sensor array, and only for the highest modes considered (modes 5 and 6), spatial resolution becomes an issue. On the other hand, MAC indices computed among mode shapes identified using the existing (and relatively sparse) CSMIP array exhibit similarity even among the lowest two mode shapes.

4 Options: Finite Element Model Updating

Once the modal data are identified, and an initial FE model is constructed, then either heuristic (i.e., trial-and-error) or systematic model updating studies can be carried out. For SCB, this said feat using either option is nearly intractable (due to the complexity of this bridge and the size of its FE model) unless a proper model-order reduction is carried out. Such a procedure not only delivers a computational tractable problem, but also allows the analyst to better understand what parameters control the model bridge's dynamic characteristics (at least those that are pertinent to the identified modes), and what parameters can be ignored (or simply slaved to other, but controlling, parameters). Model-order reduction itself is a computationally demanding task, but it can be carried out effectively and efficiently using *multivariate sampling methods* such as "Latin Hypercube Sampling" [12]. This particular method was adopted for SCB and reduced the updating parameters from its initial designation of 118 to a couple dozen, which include primarily the pile-soil stiffness parameters (for operational conditions or weak earthquakes, and separately, for strong earthquakes), kinematic interaction factors, concrete cracked stiffness reduction factors, and abutment stiffness parameters. The model updating was carried out systematically by connecting CSIBridge [13] to Matlab [14] through an Application Programming Interface, as illustrated in Fig. 8. The API allowed the use of Matlab's built-in functions and toolboxes for optimization without having to develop a separate finite element model of the bridge in Matlab. The

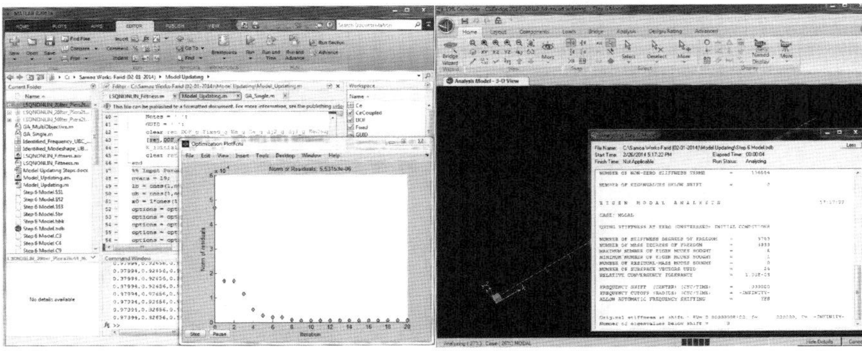

Fig. 8 Screenshot from the model updating procedure: MATLAB (*left*) and CSIBridge (*right*) are connected by an API for the updating procedures

iterative values of the updating parameters were sent by Matlab to CSIBridge, which then analyzed the structure and returned its modal properties back to Matlab. Then, the objective function—which measured the discrepancy between the identified (from recorded data) and computed (using CSIBridge) modal properties—was iteratively calculated, and the `lsqnonlin` function in Matlab was used to improve the values of the updating parameters.

Details of the model updating studies are omitted here for brevity, but may be found in [5]. The final finite element model was able to mimic the bridge responses observed during different (strong and weak) earthquakes as well as the ambient survey very well. These calibrated models provide various key observations, including the need (i) to use cracked concrete properties for the bridge superstructure that are appropriately calibrated for the anticipated level/intensity of excitation (~90 %, and ~50 % for weak and strong events, respectively), (ii) to account for kinematic interaction effects when deciding foundation input motions, (iii) to use properly calibrated soil springs, that are not necessarily in good agreement with existing predictive models that are given, for example, in [8] (even though these models offer a good initial guess, provided that pile group, diameter, and boundary condition effects can be carefully accounted for).

Representative results of the model updating studies are shown below in Table 2 and Fig. 9. As seen, an excellent agreement can be attained through carefully analyses and systematic FE model updating procedures.

5 Consequences

Once a well-calibrated FE model is obtained, then it will enable a number of calculations that can be put to good use. An example is to use to calibrated model to accurately study the interaction of the bridge with its traffic loads that affect its

Table 2 Comparison of the natural frequencies (Hz) and mode shapes (MAC) that are identified from the Ferndale 2010 earthquake, and the same computed using the final updated FE model

Mode no.	Identified frequency (Hz)	Updated FE frequency (Hz)	MAC
1	0.68	0.65	0.91
2	0.77	0.78	0.92
3	0.86	0.84	0.75
4	0.98	0.99	0.80
5	1.02	1.05	0.50
6	1.13	1.12	0.69

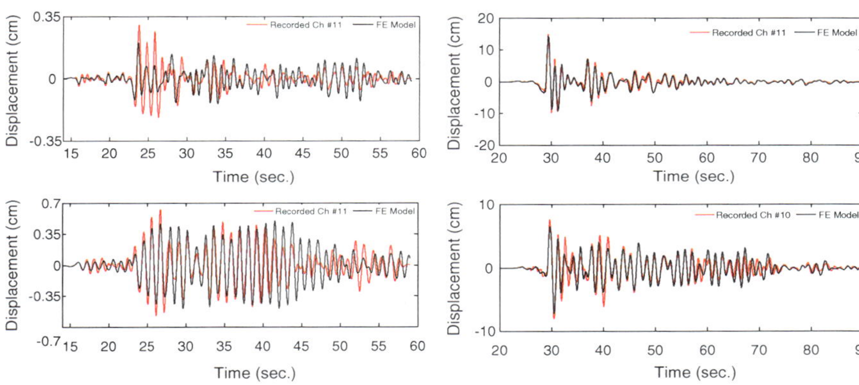

Fig. 9 Responses of SCB recorded in channels 10 and 11 during the Willow Creek 2008 (*left*), and Ferndale 2010 (*right*), along with the responses predicted for the same using the final updated FE Model

dynamic characteristics. Another important application is to carry out blind predictions of how the bridge would perform in future earthquakes. Systematically carried out, this type of simulations and parametric studies can be used to develop accurate seismic fragility curves that can be used for rapid post-event seismic safety assessment.

For SCB, an opportunity for "blind" prediction presented itself when an earthquake shook the bridge, on March 2014 that originated, yet again, in the Ferndale (offshore) area. This earthquake registered a PSA of 0.069 g (i.e., it was a relatively weak earthquake at the site of the bridge). The update FE model (without any further calibrations) was subjected to kinematic input motion obtained from the recorded pile-tip motions and the responses recorded on channels 10 and 11 are shown in Fig. 10, along with those computed using the FE model. As seen, this blind prediction yielded an excellent result.

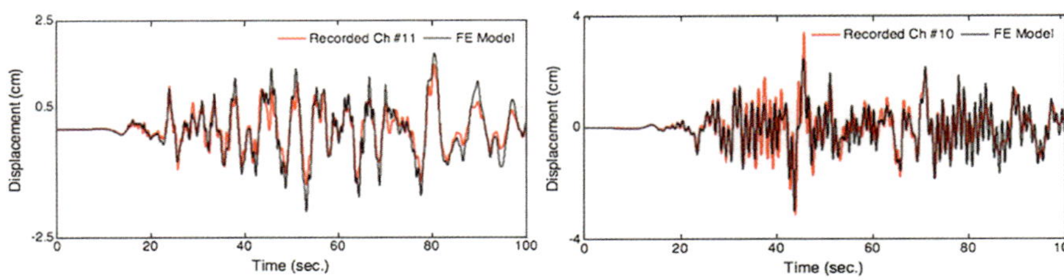

Fig. 10 The three-dimensional FE model of SCB

Acknowledgments The research presented here was funded by the California Department of Transportation (Caltrans) grant No. 65A0450. Any opinions, findings and conclusions or recommendations expressed in this paper are those of the authors and do not necessarily reflect the views of the Caltrans.

References

1. Shamsabadi A, Khalili-Tehrani P, Stewart JP, Taciroglu E (2010) Validated simulation models for lateral response of bridge abutments with typical backfills. ASCE J Bridge Eng 15(3):302–311
2. Khalili-Tehrani P, Ahlberg E, Rha CS, Stewart JP, Taciroglu E, Wallace JW (2014) Nonlinear load-deflection behavior of reinforced concrete drilled piles in stiff clay. J Geotech Geoenviron Eng ASCE 140(3):04013022
3. CSMIP. California Strong Motion Instrumentation Program, http://www.conservation.ca.gov/cgs/smip, 2014
4. Hipley P, Huang M (1997) Caltrans/CSMIP bridge strong motion instrumentation. In: Second national seismic conference on bridge and highways, Sacramento, California
5. Taciroglu E, Shamsabadi A, Abazarsa F, Nigbor RL, Ghahari SF (2014) Comparative study of model predictions and data from the Caltrans-CSMIP bridge instrumentation program: a case study on the Eureka-Samoa channel bridge. Caltrans report no. CA14-2418
6. Caltrans SDC. Caltrans seismic design criteria, v.1.7, Apr 2013
7. Matlock H (1970) Correlation for design of laterally loaded piles in soft clay. In: 2nd annual offshore technology conference 1970, Paper no. 1204
8. API—American Petroleum Institute (1993) Recommended practice and planning, designing, and constructing fixed offshore platforms—working stress design (RP 2A-WSD), Washington, D.C
9. Ghahari SF, Abazarsa F, Ghannad MA, Taciroglu E (2013) Response-only modal identification of structures using strong motion data. Earthq Eng Struct Dynam 42:1221–1242
10. Ghahari SF, Abazarsa F, Ghannad MA, Celebi M, Taciroglu E (2014) Blind modal identification of structures from spatially sparse seismic response signals. Struct Control Health Monit 21(5):649–674
11. Turek M, Ventura CE, Dowling J, Molnar S, Kaya Y (2014) Ambient vibration testing of the Eureka-Samoa channel bridge. Dyn Civil Struct 4:403–409 Springer International Publishing
12. McKay MD, Beckman RJ, Conover WJ (1979) Comparison of three methods for selecting values of input variables in the analysis of output from a computer code. Technometrics 21(2):239–245
13. Computers and Structures, Inc., CSIBridge, Version 16.0.0, 2014
14. MATLAB. The language of technical computing. Version 7.0, Mathworks, 2013

Fatigue Damage of a Diagonal Member in a Steel Truss Bridge Due to Wind-Induced Vibration

S. Nakamura, T. Okumatsu, T. Nishikawa and T. Okabayashi

Abstract Ikitsuki bridge is a three span continuous steel truss bridge completed in July, 1991. Approximately 18 years after the completion, a crack was found in a diagonal member in the vicinity of an intermediate support. In order to identify its cause, fracture surface of the crack was observed and a long-term monitoring of wind and vibration of some diagonal members has been carried out. The study reveals that (1) the crack was initiated and propagated as fatigue crack; (2) vibration of the target members is induced by the wind of 6–8 m/s and more than 15 m/s in the direction approximately normal to the bridge longitudinal axis; (3) the members mainly vibrate in the plane of truss and the vibration frequencies coincide their 1st natural frequencies; and (4) the maximum stress range induced by the vibration due to 6–8 m/s wind is approximately 30–40 N/mm^2, while that due to wind more than 15 m/s can be 195 N/mm^2. These wind-induced vibrations are thought to be the main cause of the fatigue crack.

1 Introduction

Ikitsuki bridge is a three span continuous steel truss bridge completed in July, 1991. Its side and main spans are 200 and 400 m, respectively. December, 2009, approximately 18 years after the completion, a crack was found in a diagonal member in the vicinity of an intermediate support. In order to identify the failure mode of the crack, a fracture surface observation was conducted. As a result of the observation, it was concluded that the crack was initiated and propagated as fatigue crack. Generally, the cause of fatigue crack is heavy live load or wind-induced vibration such as vortex-induced vibration. Since the bridge site is a strait between rural islands,

S. Nakamura (✉) · T. Okumatsu · T. Nishikawa · T. Okabayashi
Department of Civil and Environmental Engineering,
Nagasaki University, Nagasaki 852-8521, Japan
e-mail: shozo@nagasaki-u.ac.jp

A. Caner et al. (eds.), *Developments in International Bridge Engineering*,
Springer Tracts on Transportation and Traffic 9,
DOI 10.1007/978-3-319-19785-2_18

the traffic condition on the bridge is not severe and strong wind often blows, the wind-induced vibration was thought to be the main cause of the crack. The bridge has many members with structural and aerodynamic characteristics similar to the cracked member. Therefore, confirmation of the cause of the crack is very important to determine the appropriate countermeasure for crack prevention in other members. From this point of view, a long-term monitoring of vibration of some diagonal members and wind near them has been carried out. This paper presents the outline of the crack found and long-term monitoring to identify the cause of the crack.

2 Description of the Crack Found

A crack was found during an inspection for determining the points to be annually inspected. The member in which the crack was found is shown in Fig. 1. It is a diagonal member having a rectangular box section. The width of flange and web is 500 and 574 mm, respectively. The thickness of both plates is 9 mm. Appearance of the crack when it was found is shown in Fig. 2. The length of the crack was approximately 47 cm on the flange facing sea and 51 cm on the web facing Ikitsuki island. The maximum width of the crack was 4 mm. Almost the half of the section was fractured. The crack propagated along the weld toe on the outside of the flange facing sea.

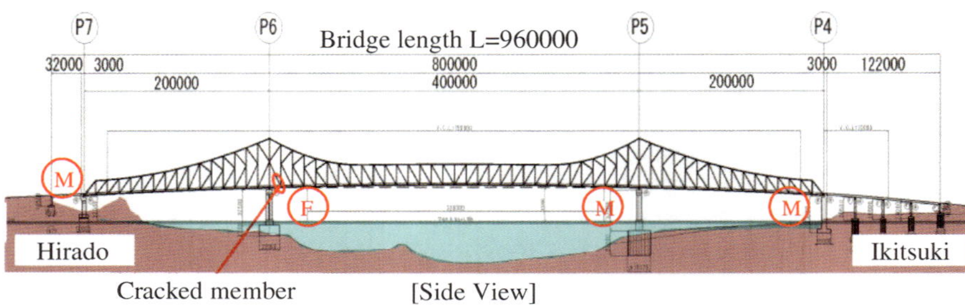

Fig. 1 Side view of Ikitsuki bridge

Fig. 2 Photos of the crack found

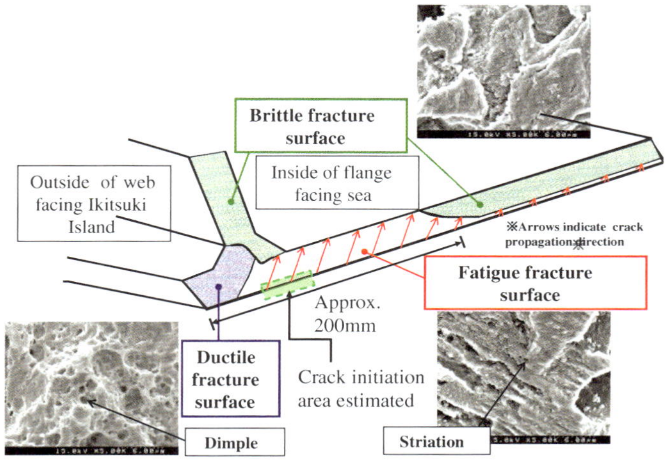

Fig. 3 Summary of crack observation result

3 Fracture Surface Observation

In the macroscopic observation of the crack surface, no typical characteristics of any failure modes such as chevron pattern and beach marks were found. Therefore, microscopic observation of the crack surface using scanning electron microscope (SEM) was conducted. Three kinds of fracture surface, i.e. fatigue fracture surface characterized by striations, brittle fracture surface and ductile fracture surface characterized by dimples, were observed. The observation result is summarized in Fig. 3. From the direction of striation, it can be thought that the crack initiated at the weld toe near the web on the outside of the flange, then, it propagated inward and widthwise as fatigue crack.

4 Long-Term Monitoring

In general, fatigue damage in bridges is induced by live load. Wind-induced vibration can be the cause as well. Since Ikitsuki bridge is located between isolated islands, the traffic condition is not very severe. Therefore, it is thought that wind-induced vibration may be the cause of the crack. In order to confirm the hypothesis, a long-term monitoring of wind and vibration of some members with similar structural characteristics to the cracked member has been carried out.

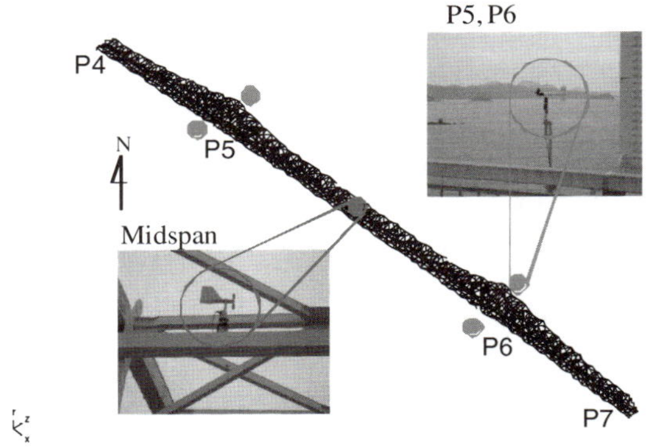

Fig. 4 Position of anemometers

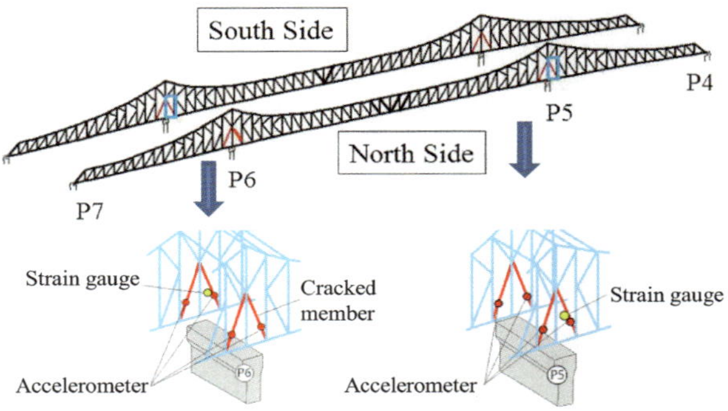

Fig. 5 Members where accelerometers and strain gauges are set

4.1 Set-up of Monitoring Equipment

Wind velocity and direction are measured by anemometers at midspan and north and south sides of two intermediate piers, P5 and P6 as illustrated in Fig. 4. Accelerometers are put on 8 diagonal members in the vicinity of intermediate support. Some strain gauges are set on two members among them. The members where the accelerometers and strain gauges are set are demonstrated in Fig. 5. Their structural characteristics are summarized in Table 1. Sampling frequency of acceleration and strain is 100 Hz. Figure 6 is photographs to show the position of strain gauges. The data obtained by P5-IN7 and P6-OS7 are mainly used in the following analysis.

Table 1 Structural characteristics of measured members

Member position		Center span	Side span
Length (m)		26.803	27.335
Section		Rectangular box	Rectangular box
Flange	Width (mm)	500	500
	Thickness (mm)	9	15
Web	Width (mm)	574	558
	Thickness (mm)	9	11
Material		JIS-SS41	JIS-SS41
Design axial force (kN)		2220	3171

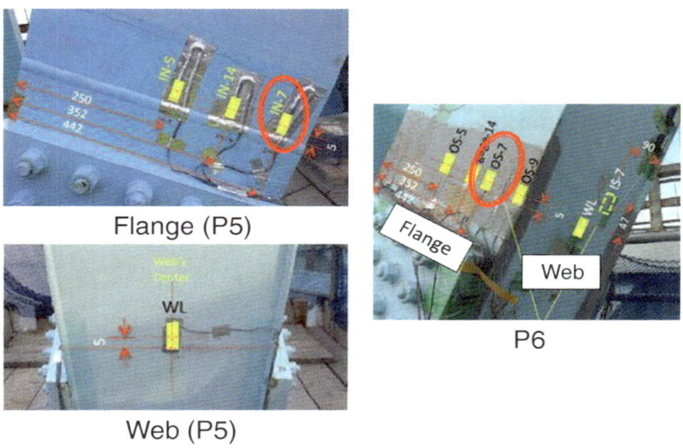

Fig. 6 Position of strain gauges

4.2 *Dynamic Characteristics of Measured Members*

Before conducting the long-term monitoring, natural frequencies and damping constant were measured by impact test. Table 2 shows its results. The 1st natural frequencies of center span and side span members are approximately 6.7 and 6.3 Hz, respectively. Comparing four members in center span, the cracked member has the frequency slightly higher than others since it was repaired. The natural frequencies of members in side span are lower than those of center span members. The damping constant is less than 0.02 for all members. By using the natural frequency and Strouhal number for rectangular section, the onset wind speed of vortex-induced vibration can be estimated as shown in Table 3. They are more than 25 m/s for both center span and side span members.

Table 2 In-plane natural frequencies and damping

Member position		P5			
		Center span		Side span	
		North side	South side	North side	South side
Frequency (Hz)	1st	6.72	6.76	6.21	6.36
	2nd	17.51	17.55	16.61	16.94
Damping constant		0.018	0.018	0.018	0.017
Member position		*P6*			
Frequency (Hz)	1st	6.83	6.74	6.29	6.29
	2nd	17.86	17.57	16.78	16.64
Damping constant		0.018	0.017	0.019	0.019

Table 3 Onset wind speed of vortex-induced vibration

Member position	Center span	Side span
D/B	0.845	0.850
Strouhal number	0.1188	0.1190
Onset wind speed (m/s)	28.3	26.1

4.3 Results and Discussions

4.3.1 Daily Maximum Stress Range

The rainflow counting method was used to obtain stress range histogram from the time history of strain record. Figure 7 shows the transition of daily maximum stress range at gauge OS-7 on a diagonal member of P6. Since the cut-off limit of Class D for variable amplitude stress in Fatigue Design Recommendations for Steel Highway Bridges [1] is approximately 40 N/mm^2, certain level of fatigue damage must be accumulated.

In Fig. 7, threshold of 30 MPa to judge the occurrence of vibration is shown. Numbers of days when the maximum stress range is more than 30 and 100 MPa are indicated in Table 4. Maximum stress range over 30 MPa was recorded in several days each year, 15 days in FY 2011, 7 days in FY 2012 and 24 times in FY 2013. Stress range more than 100 MPa was recorded only on 4 days. The recorded maximum stress range is 195 MPa in FY2011, 161 MPa in FY 2012 and 63 MPa in FY2013.

4.3.2 Wind Condition Inducing Vibration

Wind condition on the day when the daily maximum stress range is more than 30 N/mm^2 was analyzed. The analysis shows that stress range more than 50 N/mm^2 was observed under the wind with the velocity more than 15 m/s and the direction

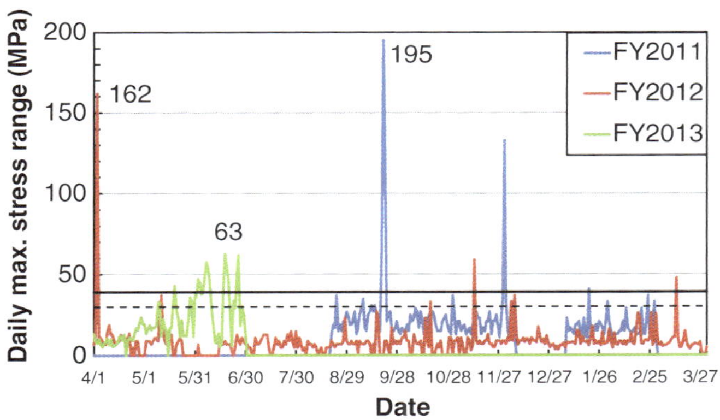

Fig. 7 Transition of daily maximum stress range

Table 4 Number of days when maximum stress range is more than 30 and 100 MPa

Fiscal year	2011	2012	2013
Over 30 MPa	15	7	24
Over 100 MPa	3	1	0

normal to the bridge axis in general. The vibration was also induced by the wind with velocity between 6 m/s and 8 m/s. In this paper, vibrations in lower and higher wind velocity ranges will be called "Phenomenon A" and "Phenomenon B", respectively.

As a typical example of Phenomenon A, relationship between maximum stress range and average wind velocity in 10 min on October 1, 2012 is shown in Fig. 8. It can be known that vibration is induced by the wind with velocity of 6–8 m/s and direction between north-northeast and northeast, approximately normal to the bridge axis. The induced stress range is less than or equal to 30 MPa. In all other cases of Phenomenon A, the maximum stress ranges were approximately 30–40 N/mm^2. The contribution of this phenomenon on fatigue damage may be small since the stress range is relatively small.

Figure 9 shows the relationship between maximum stress range and average wind velocity in 10 min on April 3, 2012. Large stress range up to 90 MPa is induced by the wind with velocity more than 15 m/s and direction between southwest and west, approximately normal to the bridge axis. But vibration was not occurred when wind direction is west-northwest, close to the direction of bridge axis even though the wind velocity is more than 15 m/s. The maximum stress range was 160 MPa although it is not shown in Fig. 9 due to the lack of the corresponding wind data. In other cases of Phenomenon B, the maximum stress range up to 195 MPa was recorded.

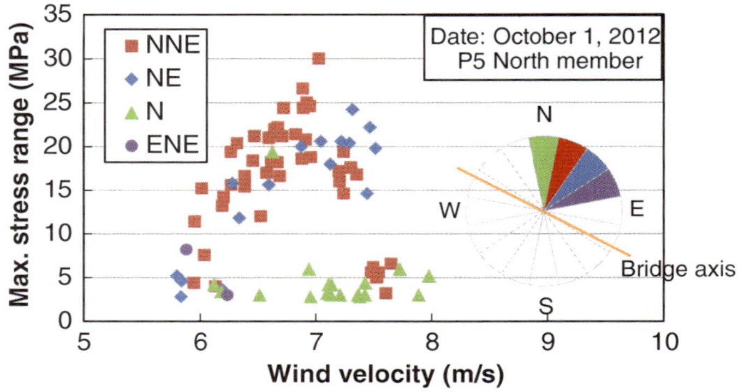

Fig. 8 Maximum stress range versus average wind velocity in 10 min (October 1, 2012)

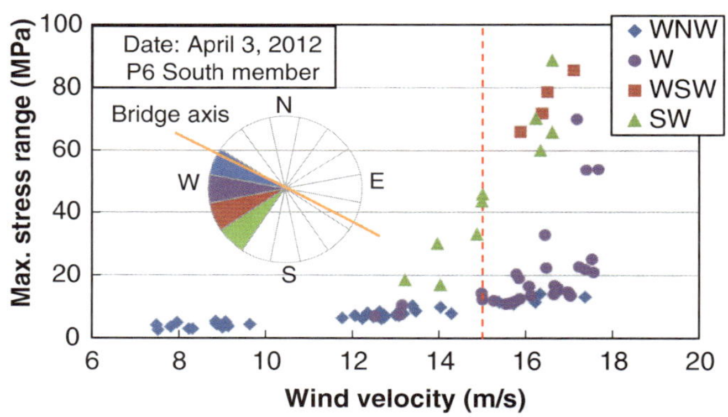

Fig. 9 Maximum stress range and average wind velocity in 10 min (April 3, 2012)

4.3.3 Time History of Stress During Wind-Induced Vibration

An example of stress time history when Phenomenon A occurs is shown in Fig. 10a. The data for 10 min was recorded at gauge IN-7 on the P5 member from 23:50, October 1, 2012. In Fig. 10b, stress time histories at gauges IN-5, IN-14 and IN-7 are compared for 4 s including the time when the maximum stress was recorded. Stress amplitude at Gauge IN-7 moderately changes in 10 min. Meanwhile, it is almost constant for 4 s. Stress at IN-5 is quite small, and stress at IN-7 is approximately twice larger than that at IN-14. Since this ratio is the same as the ratio of distances from the neutral axis for in-plane bending of the member, it can be known that the member vibrates in the plane of truss. The phases of time histories of IN-14 and IN-7 coincide and their frequency is the same as the in-plane 1st natural frequency of 6.2 Hz. Consequently, it can be said that Phenomenon A is the relatively stable in-plane harmonic vibration in their 1st natural frequency.

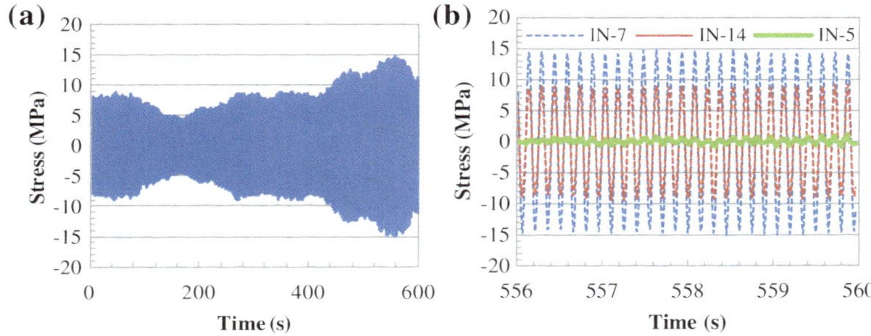

Fig. 10 Examples of stress time history when phenomenon a occurs, **a** Ten minutes history at Gauge IN-7, **b** Four seconds history at 3 gauges

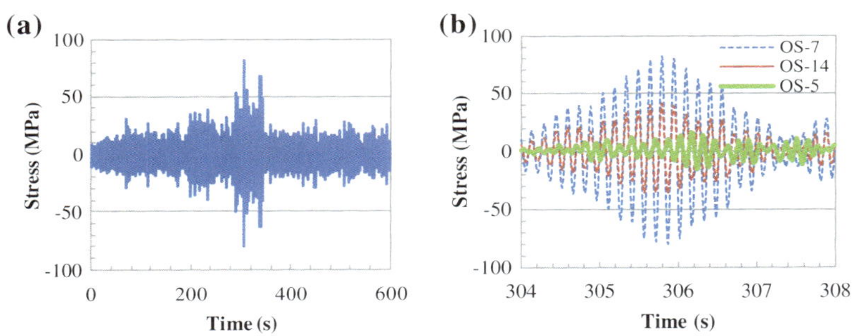

Fig. 11 Examples of stress time history when Phenomenon B occurs, **a** Ten minutes history at Gauge OS-7, **b** Four seconds history at 3 gauges

In the same way as Figs. 10 and 11a, b show the examples of stress time history when Phenomenon B occurs. The stress for 10 min in the figure was recorded from 6:50, April 3, 2012 on the P6 member. Compared with Phenomenon A, stress amplitude significantly fluctuated. The time with large stress amplitude is very short. Even for 4 s around the time when the maximum stress occurs, the stress does not keep constant. The ratio of stress between OS-14 and OS-7 indicates that the member mainly vibrates in the plane of truss at its 1st natural frequency. The stress at OS-5 and its phase difference from others suggest that a certain level of out-of-plane vibration was also induced.

4.3.4 Fatigue Damage and Life Estimated

Fatigue damage of the member was calculated by modified Miner's rule in which the cut-off limit is not taken into account. The fatigue strength of the joint was assumed as "Class D" of JRA Recommendations [1]. The stress range histogram shown in

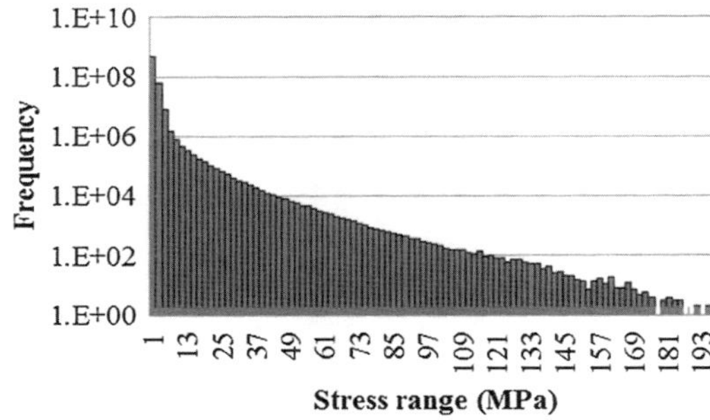

Fig. 12 Stress range histogram used for fatigue life estimation

Fig. 12 was used to calculate the fatigue damage. The histogram was obtained by applying the rain-flow method to stress at the gauge OS-7 on a diagonal member of P6 from August 2011 to December 2012. The estimated fatigue life was 57 years. It is approximately 3 times longer than the real life of 18 years. Considering that the vibration inducing fatigue damage is rare phenomenon and the fatigue strength of welded joints generally is widely scattered, the error in the fatigue life estimation may be acceptable. Therefore, it can be concluded that the main cause of the fatigue crack is the wind-induced vibration in two wind conditions.

5 Conclusions

In this study, a long-term monitoring of vibration of eight diagonal members and wind close to them was carried out to identify the cause of the fatigue crack. The monitoring reveals that (1) vibration of the target members is induced by the wind of 6–8 m/s and more than 15 m/s in the direction approximately normal to the bridge longitudinal axis; (2) the members mainly vibrate in the plane of truss and the vibration frequencies coincide their 1st natural frequencies; and (3) the maximum stress range induced by the vibration due to 6–8 m/s wind is approximately 30–40 N/mm^2, while that due to wind more than 15 m/s can be 195 N/mm^2. These wind-induced vibrations are thought to be the main cause of the fatigue crack. However, the phenomenon occurred only several times a year. Thus, accurate estimation of fatigue life by the data obtained in relatively short-term is difficult. The monitoring will be continued to obtain sufficient data for statistical life estimation.

Acknowledgments The long-term monitoring has been conducted with the support of Nagasaki Prefecture. The authors would like to express their sincere gratitude to the support.

Reference

1. Japan Road Association (2002)Fatigue design recommendations for steel highway bridges, Maruzen Co., Ltd.: Tokyo, Japan, (in Japanese)

Structural Health Monitoring of a Curved Continuous Steel Box Girder Bridge Under Marine Environment

Airong Chen, Zhen Zhang and Rujin Ma

Abstract Different from bridges on land, bridges under marine environment are subjected to tide and wave. In order to understand the actual performance of bridge under marine environment, a structural health monitoring system (SHMS) is designed and implemented on a curved continuous steel box girder bridge in the Hangzhou Bay of China. Through the implementation of the SHMS, both environmental parameters and structural response are monitored, including wind, temperature, vibration acceleration, and bearing deformation. By analyzing the monitoring data, characteristics of the environment and structural response are obtained, including the wind field characteristics, the temperature distribution of the steel box girder and the structural dynamic characteristics. From the monitoring results of the girder vibration acceleration, there is an obvious vibration phenomenon found in the lateral direction. Further studies show that the structural vibration has a direct relationship with the tides in the Hangzhou Bay. The obvious vibration is induced by regular ebb and flow, because the lateral modal frequency is as low as about 0.5 Hz which is in the range of the tidal frequencies. Moreover, the foundation scour caused by tide will lower the structural integral stiffness and then the natural frequencies, which may make matters worse. Meanwhile, finite element method is used for structural characteristics analysis and structural response analysis. Comparing the theoretical calculation results and the measured ones, the structural finite element model is verified and modified. And the modified model is used in evaluating and predicting the safety and status of the structure. In the end, some conclusions and management suggestions are given for the bridge under extreme conditions.

A. Chen (✉) · Z. Zhang · R. Ma
Department of Bridge Engineering, Tongji University, Shanghai 200092, China
e-mail: a.chen@tonji.edu.cn

A. Caner et al. (eds.), *Developments in International Bridge Engineering*,
Springer Tracts on Transportation and Traffic 9,
DOI 10.1007/978-3-319-19785-2_19

1 Introduction

With the development of calculation theories, materials and construction technologies, more and more sea-crossing bridges have been constructed or being under construction in China. Compared with bridges on land, these sea-crossing bridges are subjected to tide and wave, which may result in structural vibration and foundation scour. In these sea-crossing bridges in China, Hangzhou Bay Bridge is one of the most wonderful and famous with a total length of 36 km across the Hangzhou Bay, located in the southeast of China [1], as in Fig. 1. In this project, curved continuous steel box girder ramp bridges are designed and constructed to connect the main bridge and the offshore platform, as in Fig. 2. The main span arrangement of ramp bridges is 6 × 20 m, with single column pier of single pile, as in Fig. 3.

As a result of complex hydrological conditions of the Hangzhou Bay, the ramp bridges have a severe foundation scour problem. To makes matters worse, the land reclamation on both shores of the Hangzhou Bay will make the hydrological situation more complicated and result in more serious scour problem. Meanwhile, tide-induced vibration was found during construction and exists during operation. Therefore, a structural health monitoring system (SHMS) is specially designed and implemented on the ramp bridges for further study of the structural behavior.

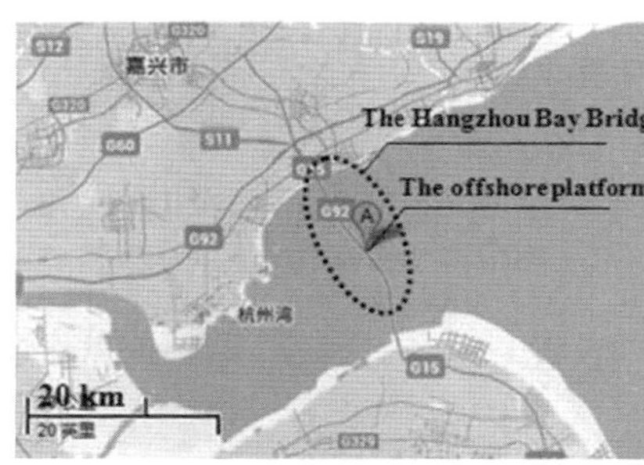

Fig. 1 Location of the Hangzhou Bay Bridge and the offshore platform

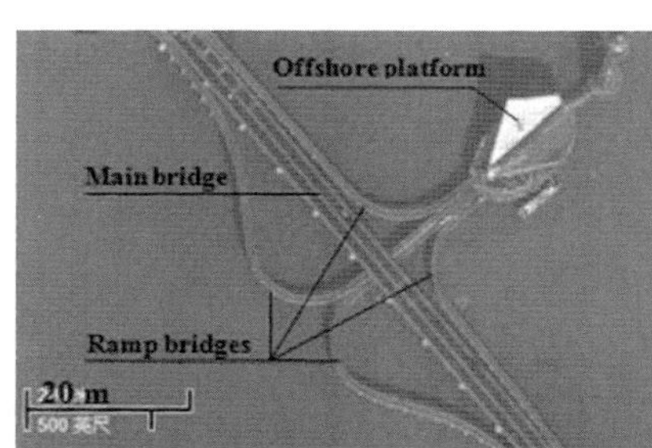

Fig. 2 Ramp bridges connecting the main bridge and the offshore platform

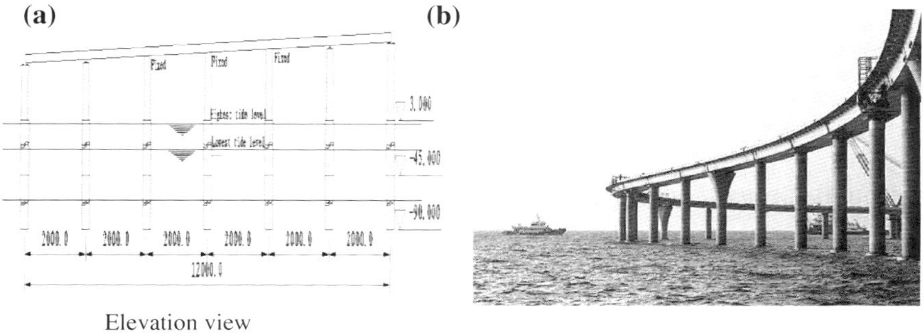

Fig. 3 The ramp bridges with single column pier of single pile, **a** elevation view, **b** side view photo

2 Design and Implement of the SHMS

In the SHMS, both environmental parameters and structural response are monitored by integration of sensor technology, measurement technology, communication technology and computer technology [2]. The SHMS consists of four subsystems, which are the sensor system, the data acquisition and transmission system, the data processing and management system, and the structural evaluation system.

The sensor system includes ultrasonic anemometer, temperature transducer, displacement transducer, accelerometer and inclinometer. Integrated circuits of low noise amplifier and filter are adopted in signal conditioning module.

Distributed data acquisition system is designed to meet the need of wide range monitoring. And those adjacent sensors are connected to the same data acquisition instrument on-site through wires. All the instruments on-site are connected to the central server by optical fiber local area network considering transmission capacity and system expansion. Data is synchronized by GPS timing module on data acquisition instruments.

The hardware part of the data processing and management system is a central server with a large capacity hard disk array, while the software part is specially developed for data collection, processing and storage. For data processing, various algorithms are used including time domain analysis, frequency domain analysis, and correlation analysis. Both original and processed data can be displayed on the monitoring interface, which is based on Web page.

The structural evaluation system is specially developed by comparing the identified results and the theoretical analyzed results. For the ramp bridges' response, both displacements of bearing and girder vibration amplitude are critical to structural evaluation.

3 Structural Performance Based on Long-Term Monitoring Data

The monitoring content includes wind, temperature, support displacement and girder vibration. There are totally 1 ultrasonic anemometer, 25 temperature transducers, 40 displacement transducers, 50 accelerometers and 18 inclinometers.

3.1 Wind Parameters

A 3D ultrasonic anemometer is installed above the deck to monitor the wind parameters, which include wind speed, wind direction, wind attack angle. Based on long-term monitoring data, the actual wind field characteristics are obtained, including distribution of wind speed and direction, turbulence intensity and gust factor. In order to study wind effect on the bridge, wind parameters during a typhoon are analyzed in detail, which happened on 8th August, 2012.

Distribution of horizontal average wind speed and direction at interval of 10 min is shown in Fig. 4a. Along wind gust factor and mean wind speed changing with time are shown in Fig. 4b. The mean wind speed during typhoon is over 10 m/s and the maximum value is about 13.9 m/s. The wind direction changes from northeast to southeast, which matches with the change of location of typhoon center relative to the bridge from southeast to southwest. The gust factor decreases below 1.5 as the wind speed increases and remains stable.

3.2 Temperature

For continuous girder bridge, expansion and contraction of main girder according to the change of temperature will lead to displacement at sliding bearings and horizontal thrust on fixing supports. As for the ramp bridge, the temperature effect is

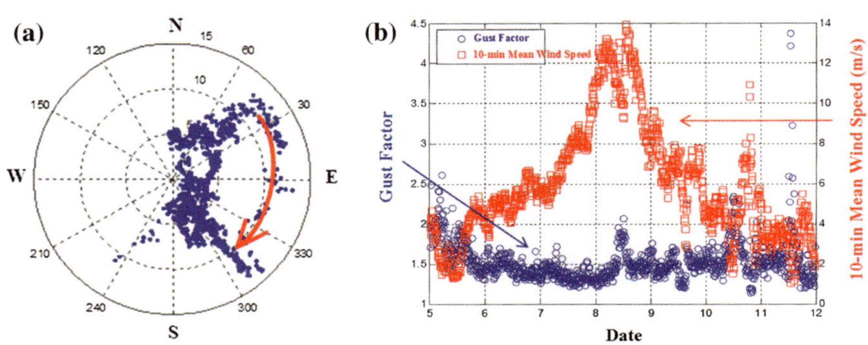

Fig. 4 Examples of the monitored wind parameters during a typhoon, **a** mean wind speed and direction map, **b** along wind gust factor and mean wind speed

more complicated and may result in disadvantageous structural response. In order to study the actual temperature effect, monitoring of temperature is carried out. Platinum resistance temperature transducers are adopted to monitor both air temperature and steel box girder temperature. Through 20 temperature transducers on a section of the box girder, the structural temperature distribution is monitored. Based on long-term monitoring, temperature distribution under different conditions are acquired and the most unfavorable distribution under high temperature is shown in Fig. 5. The most unfavorable high temperature distribution happens at 15:05 on 4th July, 2012, and the external air temperature is about 35.4 °C.

It is found that the temperature distribution of the steel box girder is sensitive to solar radiation. The temperature on top plate is obviously higher than other plates because the top plate is directly subject to solar radiation. The temperature on web plate decreases from top to bottom as a result of temperature conduction in steel material. And the temperature on bottom plate is the lowest but a little higher than the external air temperature, because it is mainly affected by temperature conduction on both internal and external boundaries of air and steel material.

3.3 *Displacement of Bearing*

As the length of continuous girder bridge increases, the total temperature deformation at both ends will certainly increase. Therefore, the bearing at both ends must function well to adapt to the requirement of temperature deformation. Otherwise, the constrained deformation will inevitably cause considerable thrust and bending moment on bridge piers. It is important to monitor the working condition of bearings. There is a pair of bearings at each end of the continuous steel box girder bridge. One is longitudinal sliding bearing, while the other is bi-directional sliding bearing. Therefore, 5 displacement transducers are installed at every bearing pair, which consists of 2 longitudinal ones, 2 vertical ones and 1 lateral one.

To study bearing deformation, monitored displacement values are analyzed with monitored air temperature. In Fig. 6, both longitudinal displacement of one sliding

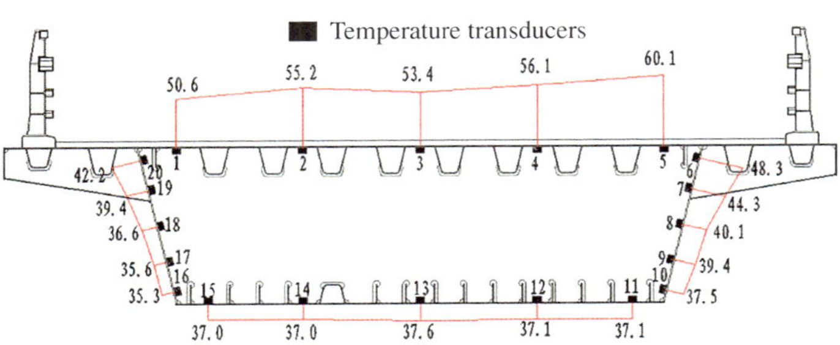

Fig. 5 Structural temperature distribution on a section of steel box girder (°C)

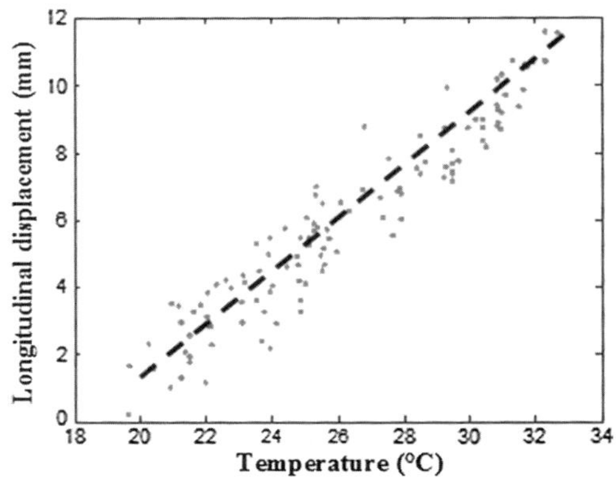

Fig. 6 Statistics of longitudinal displacement versus air temperature

bearing and the air temperature are averaged by interval of 10 min. Obviously, bearing deformation has certain linear correlation with air temperature at long time scale. Long-term monitoring data shows that the bearings function well.

3.4 Girder Vibration

Numerous accelerometers are installed on steel box girder to monitor girder's vibration in longitudinal, lateral and vertical directions. Original acceleration data shows that vehicle-induced vibration is prominent in all three directions, as in Fig. 7a. Although vehicle-induced signal is prominent in amplitude, it is high-frequency and decays rapidly. So a low pass filter is designed to eliminate the interference of vehicles. After filtering vehicle-induced vibration signal, it is found that lateral vibration is the most obvious, which is caused by tidal wave in the Hangzhou bay. There are four obvious vibration peaks corresponding to two cycles of ebb and flow in a day [3], as in Fig. 7b.

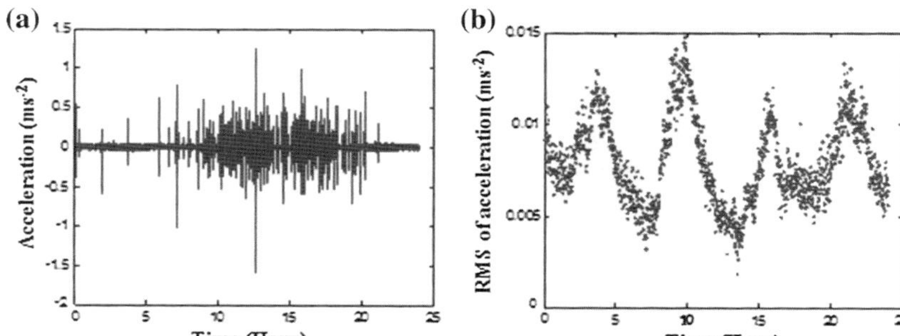

Fig. 7 Typical acceleration curve in a day, **a** original vibration data, **b** filtered vibration data

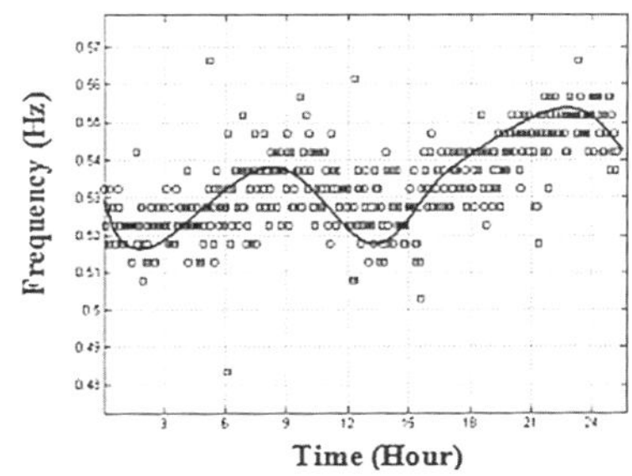

Fig. 8 Variation of dominant frequency in 24 h

Through frequency domain analysis, the structural vibration frequencies are distributed in a wide range. When looking at the dominant frequency, one interesting phenomena is observed that the dominant frequency changes in 24 h, as in Fig. 8. This frequency variation is mainly due to the change of water depth, which leads to variation of structural associated mass [4].

4 Finite Element Analysis

4.1 Finite Element (FE) Modeling and Prediction of Scouring Depth

FE method is introduced to analyze structural dynamic characteristics and to evaluate the structural status by use of the ANSYS software. Shell63 element is used to simulate the steel box girder and Solid95 element is used to simulate the concrete pier, as in Fig. 9.

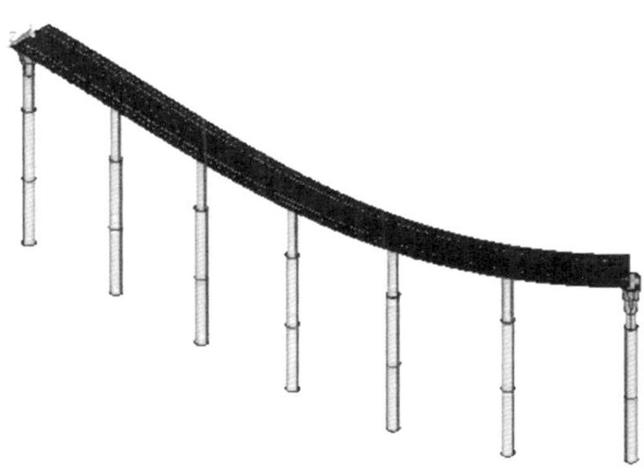

Fig. 9 FE model of curved continuous bridge

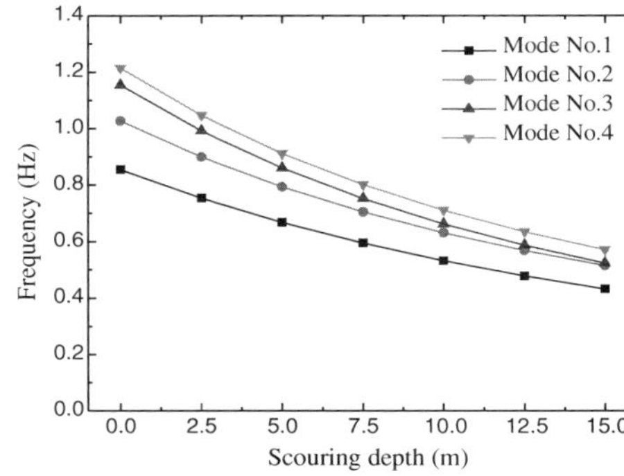

Fig. 10 Calculated basic modal frequencies versus scouring depth

4.2 FE Model Optimization

Before using the FE model, it is important to modify the original model. In this case, scouring depth is crucial because it has a great influence on the basic modal frequencies [5], as in Fig. 10. It is evident that the modal frequencies decrease as the scouring depth increasing, because the increase of free length of piers reduces the structural integral stiffness inevitably. Since material parameters and structural geometric parameters are relatively precise, measured modal frequencies and shapes can be used to determine the actual scouring depth. Based on the measured results, the current estimated scouring depth is about 13 m from completion of the structure 4 years ago. And the estimated results match quite well with the results given by annual flush detection on site.

4.3 Structural Performance Evaluation

Using the modified FE model, various calculations can be carried out to evaluate the structural status.

First of all, the temperature effect is calculated by means of FE analysis and the results are compared with the measured ones. The girder is divided into top plate, bottom plate, left web, right web and diaphragm plate. Every part is applied with actually measured temperature independently. Then displacements at both ends of the continuous bridge are obtained and compared with measured ones, as in Fig. 11. It shows that the theoretical results agree with measured ones very well. This confirms the accuracy of the FE analysis and its availability in evaluation.

Then, the effect of tide-induced vibration is evaluated by focusing on material stress according to different amplitudes, as in Fig. 12a. Stresses under other kinds of load are also calculated, such as wind, temperature and vehicle, as in Fig. 12b. The most disadvantage condition of the structure is evaluated under different load combinations.

It is found that lateral vibration will lead to slight tensile stress in local area of some piers. However, the tensile stress is still within the acceptable range.

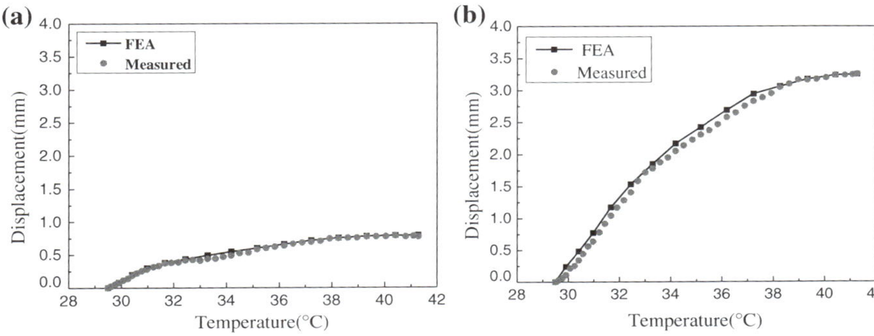

Fig. 11 Comparison of the calculated and the measured temperature displacements, **a** one end of the bridge, **b** the other end of the bridge

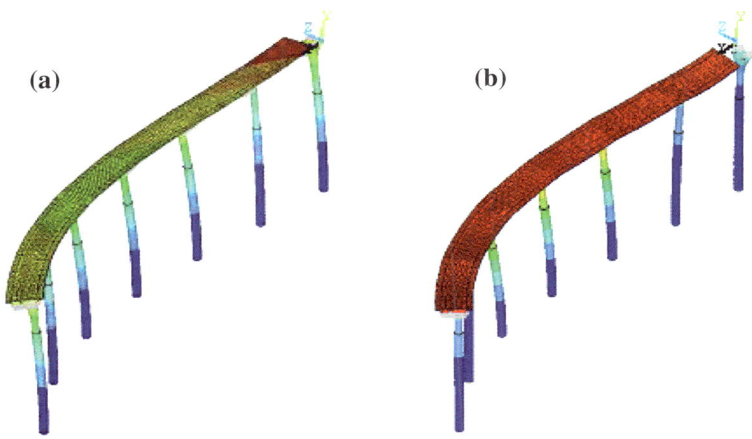

Fig. 12 Stress distribution under different loads, **a** stress under tide-induced vibration, **b** stress under vehicle load

5 Conclusions

Long term monitoring of the bridge offers the opportunity to observe structural responses under different environmental parameters. When combined with theoretical analysis, it is useful to evaluate the structural status.

- Steel box girder is sensitive to solar radiation and the difference of extreme temperature distribution is very large, which should be paid more attention during designing.
- Bearing displacements are mainly affected by overall temperature change and have a significant correlation with the temperature.

- The structural vibration has a direct relationship with regular tides in the Hangzhou Bay. The lateral modal frequency is as low as about 0.5 Hz which is in the range of the tidal frequencies. Meanwhile, foundation scour caused by tides will lower the structural integral stiffness, which may make the vibration problems worse.
- Although the ramp bridges under current condition are still in security status, continuous monitoring is suggested and more attention should be paid to foundation scour detection on site.
- Finally, the influence of tides should be taken good care of for bridges under marine environment, including foundation scouring and structural dynamic responses.

Acknowledgments The authors gratefully acknowledge the Ningbo Hangzhou Bay Bridge Development Co., Ltd. that supports the research presented in this paper.

References

1. Wang R, Meng F, Wang Z et al (2006) Overall design of the Hangzhou bay bridge. Highway 2006(09):1–7
2. Li A, Miao C, Li Z et al (2003) Health moitoring system for the Runyang Yangtze River Bridge. J Southeast Univ (Natural Science Edition) 33(5):544–548
3. Cao Y (1981) Research on tidal characteristics of the Yangtze Estuary and the Hangzhou Bay. Mar Sci 04:6–9
4. Zhu Y (1991) Wave mechanics of ocean engineering. Tianjin University Press, Tianjin
5. Chen A, Ma R, Wang D (2011) Vibration analysis of a steel curved bridge considering scouring effect. Structure engineering world conference, Lake Como, Italy

Printed by Printforce, the Netherlands